DEVELOPMENTS IN INJECTION MOULDING—1

THE DEVELOPMENTS SERIES

Developments in many fields of science and technology occur at such a pace that frequently there is a long delay before information about them becomes available and usually it is inconveniently scattered among several journals.

Developments Series books overcome these disadvantages by bringing together within one cover papers dealing with the latest trends and developments in a specific field of study and publishing them within *six months* of their being written.

Many subjects are covered by the series including food science and technology, polymer science, civil and public health engineering, pressure vessels, composite materials, concrete, building science, petroleum technology, geology, etc.

Information on other titles in the series will gladly be sent on application to the publisher.

DEVELOPMENTS IN INJECTION MOULDING—1

Edited by

A. WHELAN
M.Sc.

and

J. L. CRAFT
M.Sc., A.P.R.I.

National College of Rubber Technology,
Holloway, London, UK

APPLIED SCIENCE PUBLISHERS LTD
LONDON

APPLIED SCIENCE PUBLISHERS LTD
RIPPLE ROAD, BARKING, ESSEX, ENGLAND

British Library Cataloguing in Publication Data

Developments in injection moulding.
1
1. Injection moulding of plastics 2. Thermoplastics
I Whelan, Anthony II. Craft, J L
668.4'23 TP1150

ISBN-13: 978-94-009-9651-9 e-ISBN-13: 978-94-009-9649-6
DOI: 10.1007/978-94-009-9649-6

WITH 7 TABLES AND 135 ILLUSTRATIONS

PREFACE

Injection moulding is the most important moulding process used by the plastics industry and some idea of its importance can be obtained by considering the following figures. The value of the UK market for plastics processing equipment was £60 million in 1977. Of this sum, £23 million was spent on injection moulding machines, that is, 40 % of all the money spent on plastics processing equipment in the UK.

It has been estimated that one-third of all plastics materials are processed by injection moulding. At the present time the process is of greater importance to the thermoplastics industry but its relevance to the thermoset industry should not be ignored.

Most of the equipment now used is based on single-screw pre-plasticising units. Once these machines had become established, in the 1960s, it was felt that the ultimate had been reached in machine design and utilisation. However, since that time, machines, processes and materials have undergone extensive development to make injection moulding safer, more reliable, easier to use and more economical to operate.

The purpose of this book is to review some of the developments that have taken place in this very important area. These developments are described by specialists in the field, who have extensive industrial experience and whose contribution will therefore be of immediate relevance to those concerned with the usage and application of this, the most important plastics moulding process.

CONTENTS

LIST OF CONTRIBUTORS

C. D. BARNETT D.M.S., M.I.P.M., A.M.B.I.M., A.M.P.R.I.

Senior Training Adviser, Rubber and Plastics Processing Industry Training Board, Brent House, 950 *Great West Road, Brentford TW*8 9*ES, Middlesex, UK.*

H. BLÜML Ing. (grad.)

Development Engineer, Demag Kunstofftechnik, D-8500 *Nurnberg, Rennweg* 37, *West Germany.*

G. CHEATER C. Eng., M.I.Mech.E., M.I.Prod.E., M.I.M.C.

*Head of Productivity Management Group, Rubber and Plastics Research Association, Shawbury, Shrewsbury SY*4 4*NR, UK.*

F. COOKE

Technical Director, Optimum Evaluation and Instruction, 3 *Devon Street, London SE*15 1*JR, UK.*

P. H. J. INGHAM B.Sc., C.Eng., M.I.E.E.

*Marketing Manager, Eurotherm Ltd, Worthing, Sussex BN*14 8*NW, UK.*

R. A. IRELAND F.P.R.I.

*Technical Director, Ketch Plastics Ltd, Eastern Avenue Works, Lichfield, Staffs. WS*13 7*SD, UK.*

R. C. McGREGOR B.Sc., Ph.D., A.R.C.S.T.

Technical Adviser, Marketing Services Department, ICI Plastics Division, Bessemer Road, Welwyn Garden City, Herts. AL7 1HD, UK.

J. NIGHTINGALE B.Sc., A.R.I.C.

Plastics Consultant, 6 *High Grove, Welwyn Garden City, Herts. AL*8 *7DW, UK.*

J. B. SMITH

Account Manager, Plastics Products and Resins Department, Du Pont (UK) Ltd, Du Pont House, 18 *Breams Buildings, Fetter Lane, London EC4A* 1*HT, UK.*

G. F. TURNBULL M.Sc., Ph.D.

*Joint Managing Director, Turnbull Control Systems Ltd, Mulberry Lane, Goring by Sea, Worthing, Sussex BN*12 4*RB, UK.*

Chapter 1

DIGITAL HYDRAULICS

H. BLÜML
Development Engineer, Demag Kunstofftechnik, Nurnberg, West Germany

SUMMARY

The production cycle of an injection moulding machine, with its rapidly changing hydraulic energy requirements, calls for the use of the most modern industrial hydraulic equipment. In recent years a number of systems have become available which aim at improving the moulding process so as to give higher levels of automation and improved part quality. However, some of these systems are expensive and difficult to use.

A hydraulic system that provides ease of setting and gives good reproducibility (and has become known as digital hydraulics) will be described in this chapter.

1 INTRODUCTION

The production cycle of an injection moulding machine with its rapidly changing hydraulic energy requirements calls for the use of the most modern industrial hydraulic equipment. In this type of machine, production and process requirements must be met using the least possible space, and the machines must also be capable of being serviced easily. In addition they must be environmentally acceptable and economical in operation.

In contrast to continuous processes, which can often be optimised with the aid of fairly simple control systems, discontinuous production processes require the use of complicated and costly control techniques. At the beginning of the 1970s, a number of stratagems were known which helped to raise the injection moulding process to a higher level of automation and improved the quality of the product. In particular, the following had become necessary or available:

(i) devices which enabled individual process parameters to be adjusted or controlled upon the occurrence of interference factors;
(ii) mechanical or electrical storage of machine parameters, found to have been the optimum, which enabled such machine settings to be re-established in a reproducible manner;
(iii) the use of process computers as automatic start-up aids and for optimising production;
(iv) variation of individual parameters (e.g. injection speed and pressure) dependent upon distance or time, during the process cycle, as a method of effecting control and adjustment; and
(v) elimination of interference factors which affect the mechanical, hydraulic and electrical settings, i.e. improvements in the setting accuracy and reproducibility of a machine.

Whereas some of the above-mentioned stratagems can be practised only with great difficulty, improvements in machine control are acquiring increasing importance. Apart from the achievement of greater capacities (e.g. higher shot weights) and higher running speeds, successful efforts have been made to eliminate interferences that adversely affect the running of the machine on a reproducible basis. The most economical—though not always the most scientific—approach is by way of a fundamental analysis of the interference factors so as to eliminate the causes of interference at source. The choice of the methods applied and their quality and robustness determine the success of this stratagem.

A hydraulic system that provides ease of setting and gives good reproducibility (which has become known as digital hydraulics) will be described in later sections. However, let us first consider some of the various hydraulic systems that have been used with injection moulding machines.

2 CONTROL OF HYDRAULICALLY OPERATED INJECTION MOULDING MACHINES

Many different types of hydraulic component are available and such components may be connected together in numerous ways. Some of the systems used on injection moulding machines will be illustrated by reference to block circuit diagrams.

As initially stated, economics plays a big part in the choice of a particular system but other factors (such as, for example, the need for independent operation of some machine functions, machine noise, low installed power, etc.), must also be considered before a particular system is chosen.[1]

2.1 Single Pump Machines

The block circuit diagrams in Fig. 1 illustrate the various items which are necessary to establish the control chain for pressure, direction and speed

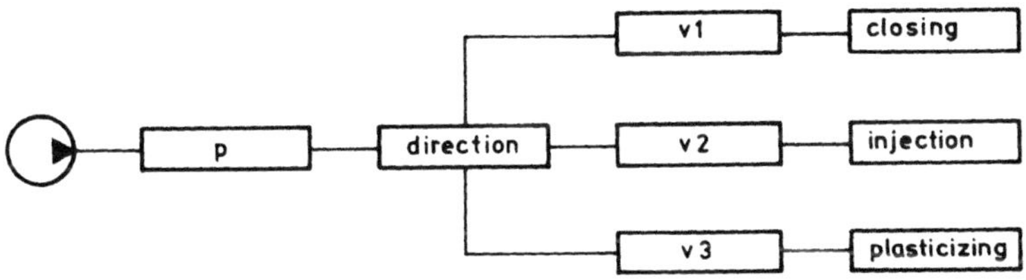

FIG. 1(a). Conventional circuit with separate means of speed control.

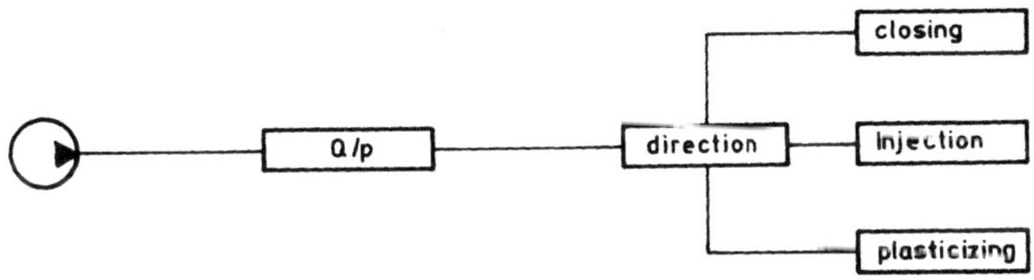

FIG. 1(b). Programmable circuit using a pressure–volume (Q/p) control device.

when the system is operated by means of only one fixed-volume pump. The usual sequence is illustrated in Fig. 1(a). The block marked 'p' represents a pressure control device, that marked 'direction' represents a switching unit or valve and those marked 'v' represent speed or volume control devices. For controlling speed, use is made of throttle valves, and, when the requirements are stringent, flow control valves are used.

If speed and pressure are to be programmable, a unit (marked Q/p in Fig. 1) for electrically controlling the amount and pressure of the oil delivered is arranged in front of the direction control device as shown in Fig. 1(b). This results in a very compact control system.

2.2 Machines with Accumulators

The pump capacity to be installed can be reduced with the aid of hydraulic

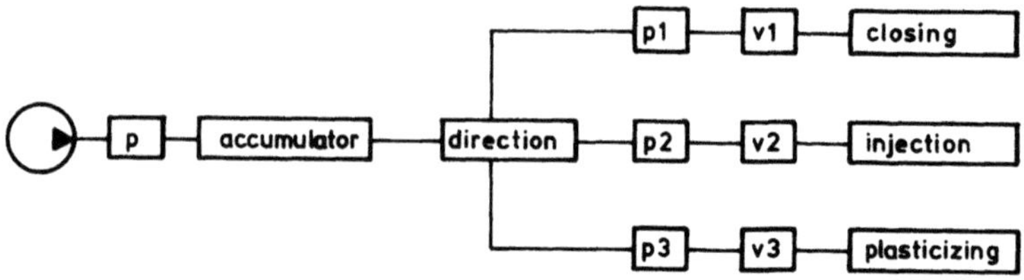

FIG. 2(a). Block diagram of single pump, accumulator circuit.

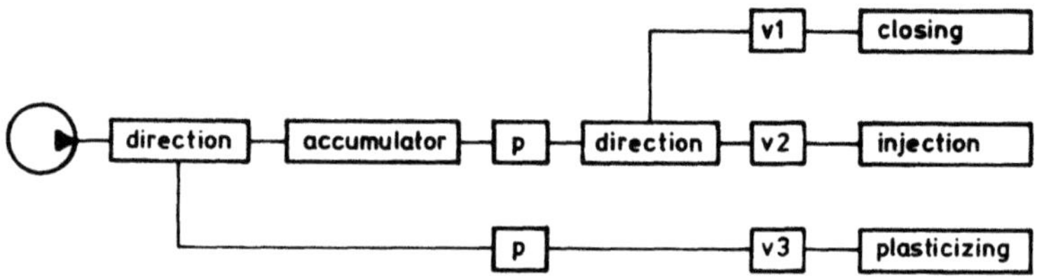

FIG. 2(b). A single pump curcuit where only the maximum element units are connected to the accumulator.

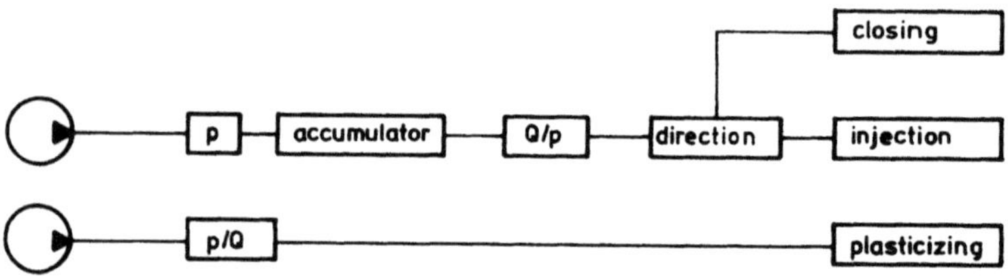

FIG. 2(c). Accumulator circuits with pressure–volume control device; plasticising with separate pump.

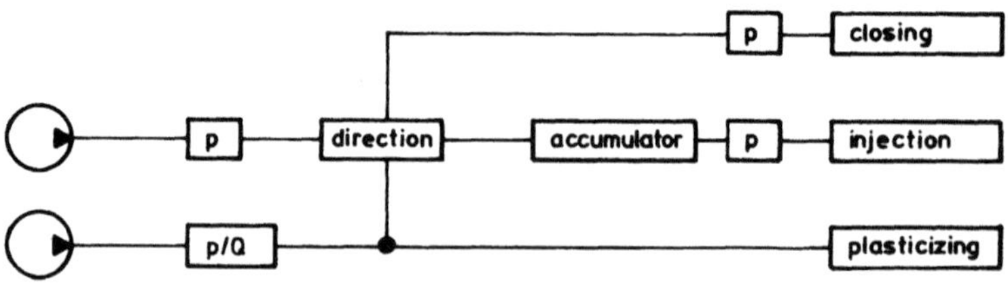

FIG. 2(d). Accumulator circuit available only for the injection stroke; plasticising with separate pump.

accumulators; this may be desirable if the hydraulic consumer units have greatly different requirements and if sufficient charging time is available.

With accumulator operation, as illustrated in Fig. 2(a), a difference occurs between the pressure built up in the accumulator and the particular working resistance that is required. The product of the pressure difference and the delivered flow of oil is a measure of the power loss caused by the accumulator. To keep power losses low, consumer units having a uniform oil requirement should be supplied directly from a pump. Proposal 2(b) shows how pressure accumulators are used for supplying those hydraulic units which have peak requirements. This is often only necessary in the injection operation.

The block circuit diagrams in Figs. 2(c) and (d) represent solutions using an additional pump for the hydraulic screw drive; pressure and speed are programmable for the hydraulic screw drive in the plasticising operation.

2.3 Multiple Pump Machines

The block circuit diagrams in Fig. 3 illustrate various means of controlling a system with multiple pumps. Solution 3(a) is the most frequently used scheme and is characterised by two pump circuits. Although one of these is,

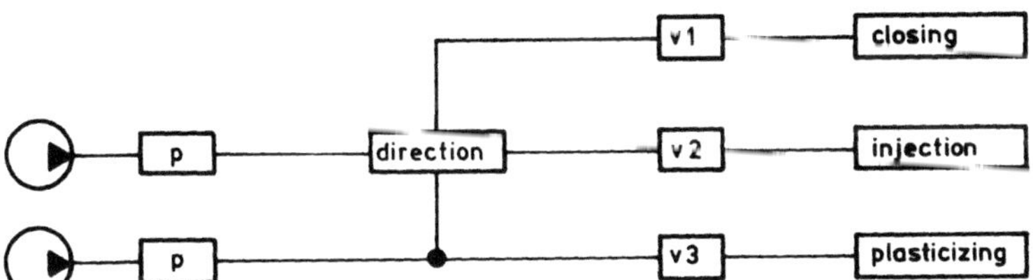

FIG. 3(a). Circuit with double pumps.

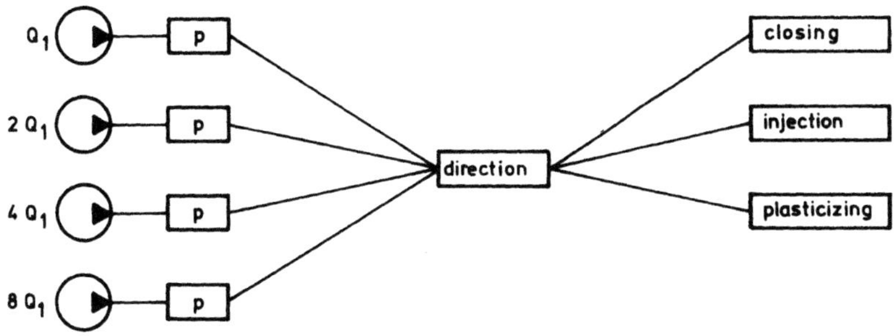

FIG. 3(b). Control of speed by pump combination.

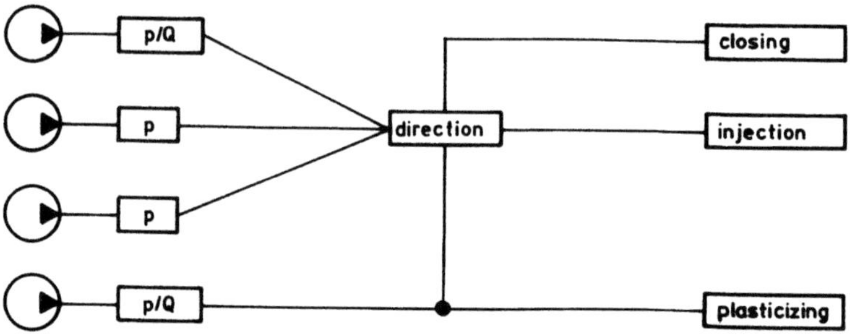

FIG. 3(c). Multiple pump system with separate pump for plasticising.

however, primarily used for supplying the hydraulic screw drive (plasticising) during injection, it is also connected up to the other pump to increase the injection speed.

With the aid of geometrically stepped delivered quantities from several fixed-volume pumps, speed can be controlled very economically. The example illustrated in Fig. 3(b) shows how, by means of electric controls, four different pumps, each giving a fixed volume, can be combined to achieve 15 different evenly stepped speeds. Naturally, as shown in Fig. 3(c), one of the pumps may also be advantageously used for independent plasticising. Pressure and speed are then again controlled directly at the pump by means of a suitable programmable valve.

2.4 Variable Volume Pumps

Adjustable pumps delivering variable volumes (on the left in Fig. 4) can be infinitely adjusted to suit the oil requirements during the cycle. Thus, control possibilities are provided that will acquire more importance in the future. Various solutions are feasible for the programming of pressure and speed. In Fig. 4(a) pressure is limited directly at the pump, whereas speed

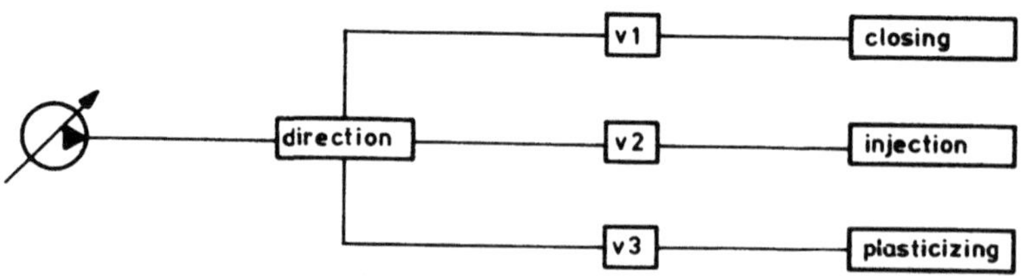

FIG. 4(a). Block circuit diagram using a variable volume pump with pressure programming.

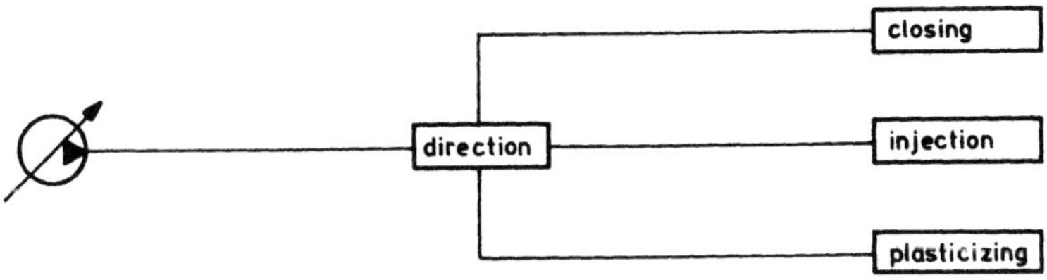

FIG. 4(b). Variable pump with speed and pressure programming.

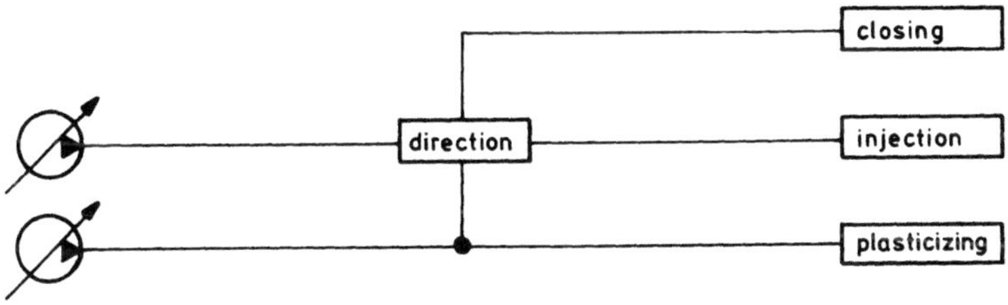

FIG. 4(c). Double pump with pressure and volume programming.

(e.g. injection speed) is set with the aid of throttle valves or volume regulators. The automatic regulation of the pump, when the required speed and the pre-programmed pressure are reached, enables the power requirements to be reduced compared with those occurring when fixed delivery pumps are used. In the variant shown in Fig. 4(b), speed programming is carried out directly at the pump. Manually adjustable throttle valves are replaced by electrically set controls. In this system priority is given to the speed setting, so that the pump only has to deliver against the particular operating pressure. This system cannot be bettered as regards power requirements. For the sake of completeness, Fig. 4(c) again shows the extension of the drive for independent plasticising using a further regulating pump.

3 METHODS OF IMPROVING MACHINE PERFORMANCE

The development of hydraulic control systems for injection moulding machines is characterised by a trend towards progressively better performance. Such a trend is accompanied by requirements for smaller sized units, smoother running systems, an increased number of functions, and greater operational reliability.

3.1 Machine Setting

The setting of electrically controllable values such as times, temperatures and optional functions is carried out at the control panel; distances, speeds and pressures are adjusted manually on the machine. Today, there are machines on the market in which pressures and speeds can also be set electrically on the control panel. In this connection it is important to realise that because the process parameters can be influenced electrically it does not always automatically follow that improved reproducibility results.[2] Certain variants of this technique simply result in greater convenience in operating the machines, and the stability and reproducibility of the adjusted hydraulic parameters do not equal those of existing conventional control systems where pressures and speeds are set manually.

3.2 Process Control Systems

To obtain a product which is uniform within narrow tolerances, a production cycle that is unaffected by external interference factors is necessary. This can be achieved by the use of electro-hydraulic control circuits consisting of a servo valve, an actual value sensor (or transducer), and the associated control electronics. Such systems, commonly known as process control systems, take care of the injection operation in particular and, within certain limits, they are capable of offsetting the effects of temperature fluctuations e.g. fluctuation of tool, melt and oil temperatures. However, a measurable increase in quality can also be achieved in machines using simple conventional control systems.

Practice has shown that the complicated techniques involved in process control systems—contrary to the reasons often put forward in their favour—require operating personnel who are trained in control techniques and in the operation of the production process. These requirements are not met in the great majority of companies concerned with injection moulding. Most processing firms require machines having hydraulic and electrical control systems that are capable of eliminating interference factors which might influence the reproducibility of the production process; in this way they avoid the need for using complex control systems.

3.3 Control of Oil Temperature

In addition to the temperature of the tool, the temperature of the oil of the hydraulic system has a decisive influence on the reproducibility of the production cycle. If the oil temperature and therefore its viscosity (Fig. 5) are kept constant, then conventional hydraulic equipment operates in an astonishingly constant manner over lengthy production periods.

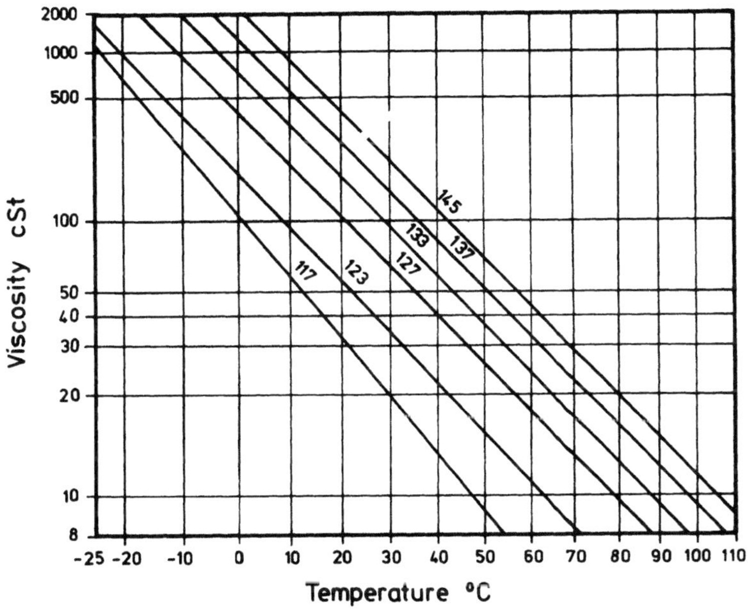

FIG. 5. Viscosity–temperature graph for some hydraulic oils.

Preheating of the oil drastically shortens the start-up phase, so that correction of the machine settings becomes unnecessary.

Figure 6 illustrates one form of preheating and control circuit for the oil of the hydraulic system. The oil is heated up with the aid of hydraulic pumps and pressure valves not illustrated in the figure. A heat sensor communicates the actual value of the temperature to an electronic on–off controller which in turn acts on a (digital) water valve. In this way a variation in oil temperature is achieved that oscillates between two very close limiting values and is independent of the machine cycle that has been set (Fig. 7).

3.4 Digital Systems and Cartridges

With the aid of new digital systems for controlling pressure and speed it is possible to achieve ease of operation of the machine together with high setting precision.

With increasingly higher operating speeds more attention is again being directed towards the switching times of hydraulic multi-way valves. This

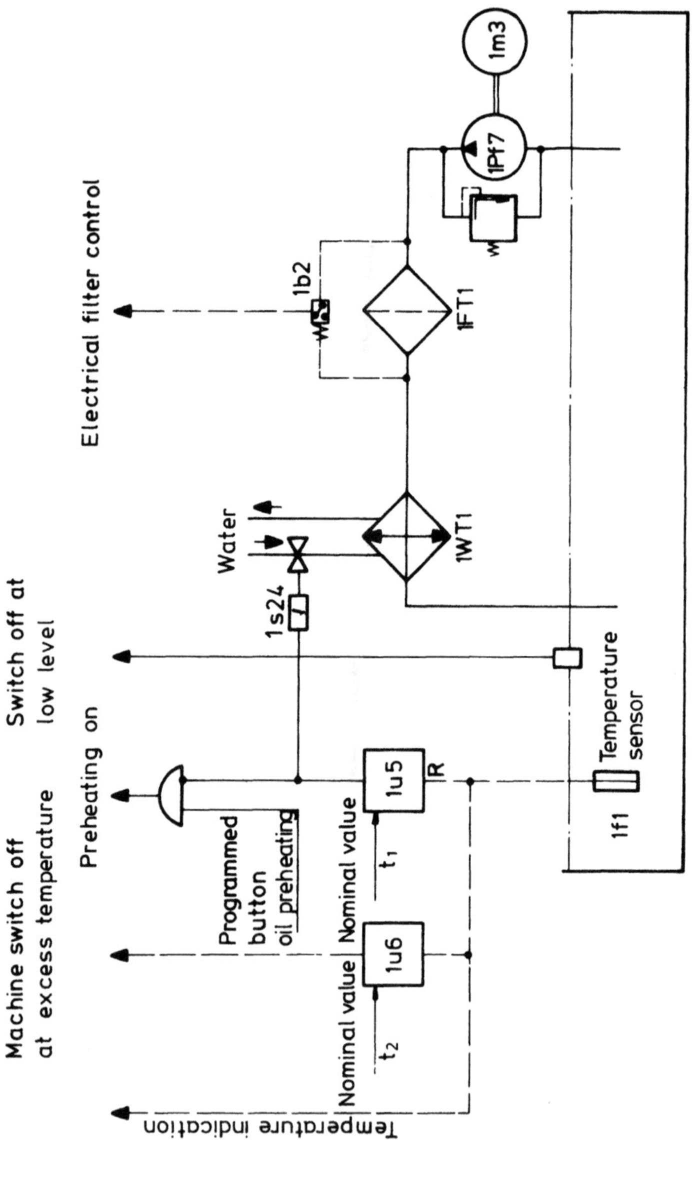

FIG. 6. Preheating and temperature regulation circuit for the hydraulic oil supply of a modern injection moulding machine.

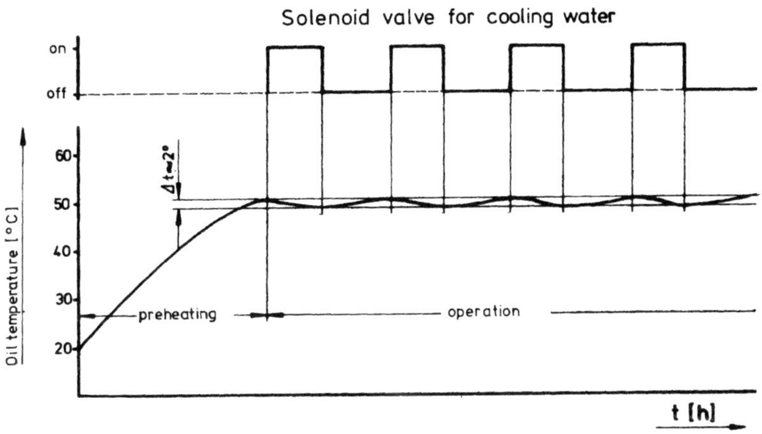

FIG. 7. Typical record of oil temperature from a machine fitted with preheating and temperature control.

is necessary to reduce (i) switching times, and thus limit their tolerances with the object of achieving improvements in performance and accuracy; and (ii) switching noise. Considerable progress has been achieved using cartridges (see Section 4.3) instead of conventional multi-way valves. Distance is controlled more precisely, rapidly and quietly, and an appreciable reduction in the size of the control equipment is achieved.

4 DIGITALLY CONTROLLED MACHINES

In all those branches of industry in which reliable transmission of data and high reproducibility are required, the digital technique is acquiring ever-increasing importance. It is an obvious step also to apply this technology to the control of injection moulding machines.

The use of conventional components (involving no greater supervision or maintenance than that used for more traditional injection moulding machines) and the simple logical function of digital elements, render this control system particularly acceptable from an operational point of view. Of the basic solutions illustrated in Figs. 1 to 4, all proposals having central pressure and volume programming are in principle suitable for electronic digital control of these parameters. The digitalised path control system constituted by the cartridge technique is advantageous from the technical

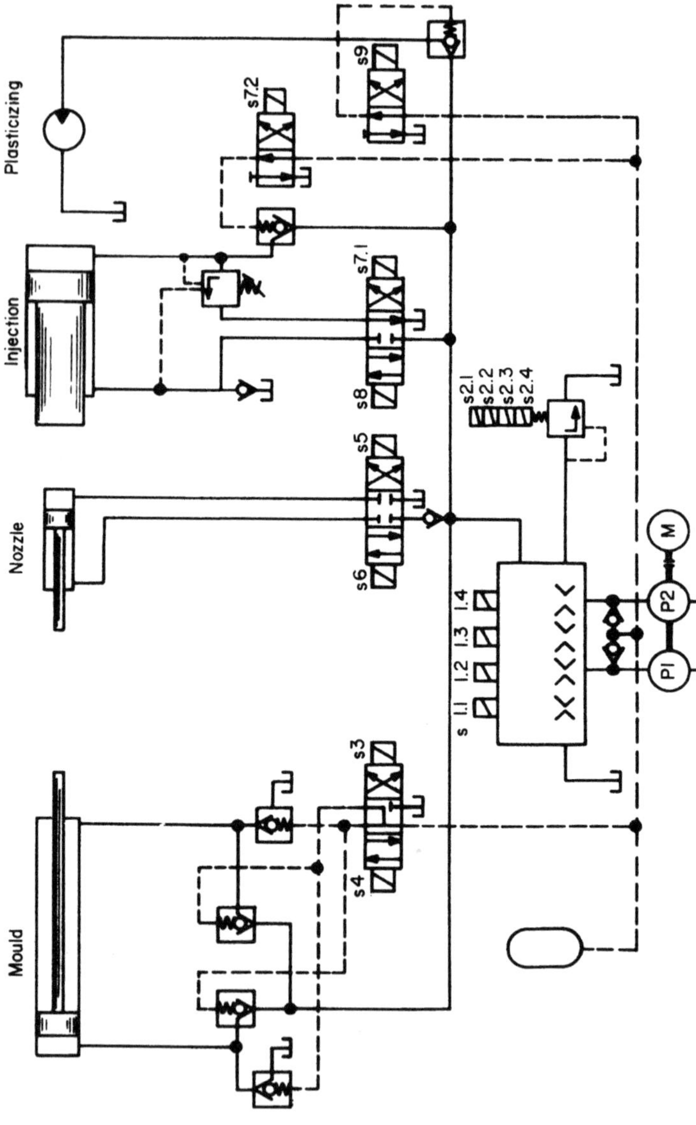

FIG. 8. Hydraulic control circuit for digital injection moulding machine.

and economic points of view when oil flowing at a rate above approximately 80 litres/min has to be controlled. (This is explained in more detail in Section 4.3.)

Figure 8 illustrates in a simplified manner the hydraulic control system of an injection moulding machine with central digital pressure and volume programming as well as digital path control by means of cartridge elements.

4.1 Speed or Volume Control

4.1.1 *The Use of Predetermined Settings*

In contrast to conventional control systems, which incorporate manually operated infinitely adjustable throttle valves or volume regulators, digital volume regulators operate with individual predetermined settings. The choice of the setting range depends on what is required and on the permitted expenditure.

Figure 9 illustrates, in principle, equipment which would give 31 evenly spaced quantities of oil. Such a volume regulator consists of a combination of five different apertures the dimensions of which have been chosen so that known volumetric quantities of oil will pass through them.

To keep these flow quantities constant even when the pressure changes upstream and downstream of the equipment, the system incorporates a differential pressure regulator, which ensures that the pressure difference at the measuring apertures is constant. If required, it is possible to arrange the control elements in series or parallel with the measuring apertures, and we thus obtain either two-way volume regulators or, as shown in Fig. 9, energy-saving three-way volume regulators. The apertures are so dimensioned that

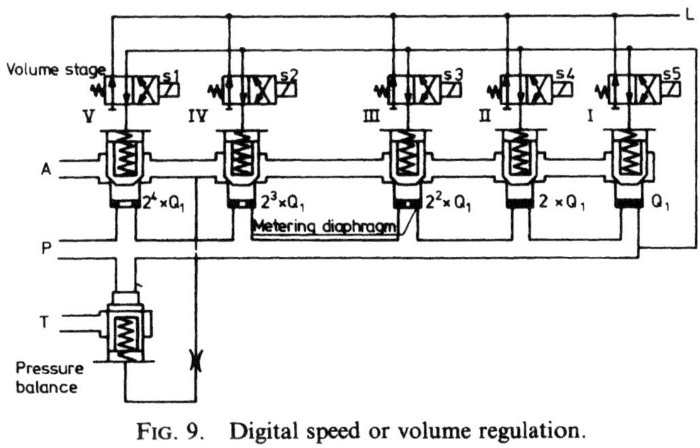

FIG. 9. Digital speed or volume regulation.

Step	Code				
	s1	s2	s3	s4	s5
0					
1	×				
2		×			
3	×	×			
4			×		
5	×		×		
6		×	×		
7	×	×	×		
8				×	
9	×			×	
10		×		×	
11	×	×		×	
12			×	×	
13	×		×	×	
14		×	×	×	
15	×	×	×	×	
16					×
17	×				×
18		×			×
19	×	×			×
20			×		×
21	×		×		×
22		×	×		×
23	×	×	×		×
24				×	×
25	×			×	×
26		×		×	×
27	×	×		×	×
28			×	×	×
29	×		×	×	×
30		×	×	×	×
31	×	×	×	×	×

FIG. 10. The ways in which flow rates can be combined to give a range of speeds.

the streams flowing through them behave like a geometric series, their relationship being, for example, 1:2:4:8:16. Each stream can be brought into action by way of path control valves and cartridges. Because of the binary gradation, when 'n' measuring apertures are used, $z = 2^n - 1$ volume settings can generally be obtained. In our example, 31 speed values are obtained with five apertures. Figure 10 shows that by using different combinations of the five apertures 31 speed settings can be achieved. The reproducibility of the setting that is so important in injection moulding is ensured by the following features:

(a) The required quantity of oil is achieved by electrical signals which are set digitally on a central control panel. Within the permissible voltage tolerances for multi-way valves, the signals are transmitted without variation when such centralised setting is employed.
(b) The apertures through which the oil flows are preferably circular and their measuring function is thus better than that of the adjustable orifices in volume regulators.

A considerable advantage of oil-volume regulation using digital volume regulators is achieved in cases in which tools are frequently changed. Because of the close manufacturing tolerances of the orifices, as well as of the springs of the differential pressure regulator, values set on the digital switch can easily be reproduced precisely, not only on one machine but also on other machines of the same type and capacity.

Volume regulators are purely hydraulic devices for controlling flow volume. If leakages in the downstream control components, cylinders and motors are negligible—and this is so in the vast majority of cases—then speed is proportional to the volume of oil.

To achieve stepped speed profiles, use can likewise be made of digitally controllable volume regulators. On the other hand, digital hydraulic devices having finite resolution are not suitable as the adjusting elements in an electro-hydraulic control circuit.

Apart from the above-described digital volume regulator, other systems for controlling speed by means of pre-programmed partial volumetric flows have become known, and these are specified below for the sake of completeness.

4.1.2 *Other Types of Digital Volume Regulator*

The adjusting stroke of a normal volume regulator occurs in stages with the aid of a positioning cylinder. The stepped flow characteristic curve depends upon the form of the control orifices and may be both linear and progressive.

Another type of digital flow device that has been tried out on a practical basis makes use of several pumps, each delivering a different quantity of oil. The output of these fixed volume pumps is geometrically stepped (see Fig. 11) so that operating speed requirements may be met by combining the outputs in various ways.

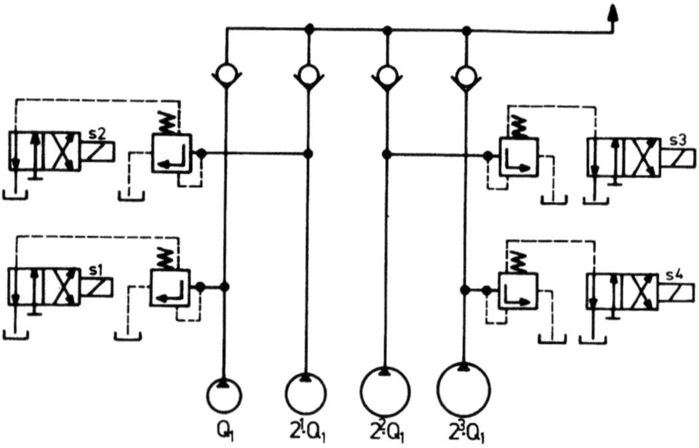

FIG. 11. Speed control by combining the output of four pumps.

It is an obvious step also to control the output of adjustable pumps by means of digital control elements. Speed setting, by varying the quantity delivered by the pump, is the most economical way of utilising power. However, it widens the pressure-dependent tolerance range by reducing the speed of the motor driving the pump. The volume of oil obtained is decreased and the pressure obtained is not so consistent.

4.2 Digital Pressure Regulation

Digitally adjustable pressure-limiting and pressure-regulating valves are advantageously used instead of analogue-adjustable valves when resolution is secondary to reproducibility. Good reproducibility is necessary in injection moulding because of the discontinuous mode of operation. As in digital volume control there are variations on the theme of digital pressure control.

4.2.1 *Digital Setting with Analogue Control*

A common feature of some digital pressure control systems is the transmission of digital signals from the setting point to a hydraulic valve. This valve constitutes a digital-to-analogue converter and converts the incoming digital signal into a control pressure which is used for limiting and regulating the line pressure (Fig. 12) in a conventional way.

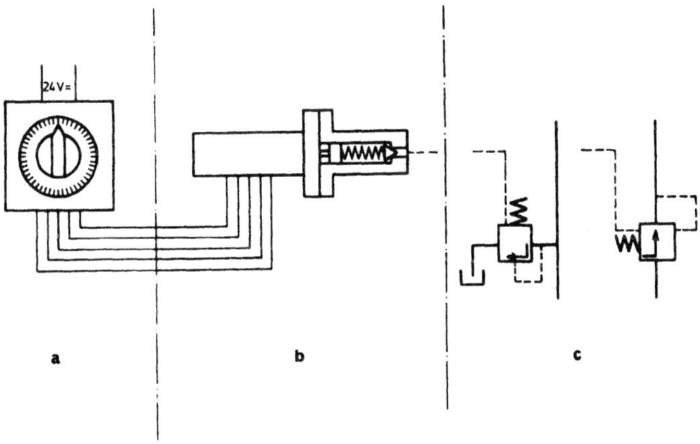

FIG. 12. Use of a digitally set pilot valve to control pressure. (a) Step switch with $2^n - 1$ settings in the control cabinet. (b) Conversion of digital signals to control pressure signals. (c) Main valve setting (in the machine) for pressure limitation or control.

Pressure valve with digital cylinder. The device shown in Figs. 13 and 14 is a combination of a pressure setting unit (remotely set) and an adjustable cylinder (which gives a control pressure); this device replaces the manual setting system commonly found on machines. The final control drive from the adjustable cylinder consists of small cylinders arranged in series, the strokes of which are precisely limited by mechanical stops. Each stroke is different and behaves in accordance with the binary number code. The stroke of the final positioning piston is made up of the sum of the controlled individual strokes. By means of a compression spring with a flat characteristic curve the stroke is first converted into a force and this in turn is converted by way of the valve cone into the required control pressure. In this way the attainment of accurate setting reproducibility, which is of decisive importance to production tolerances in injection moulding, is ensured. Digital signal independence (of both temperature

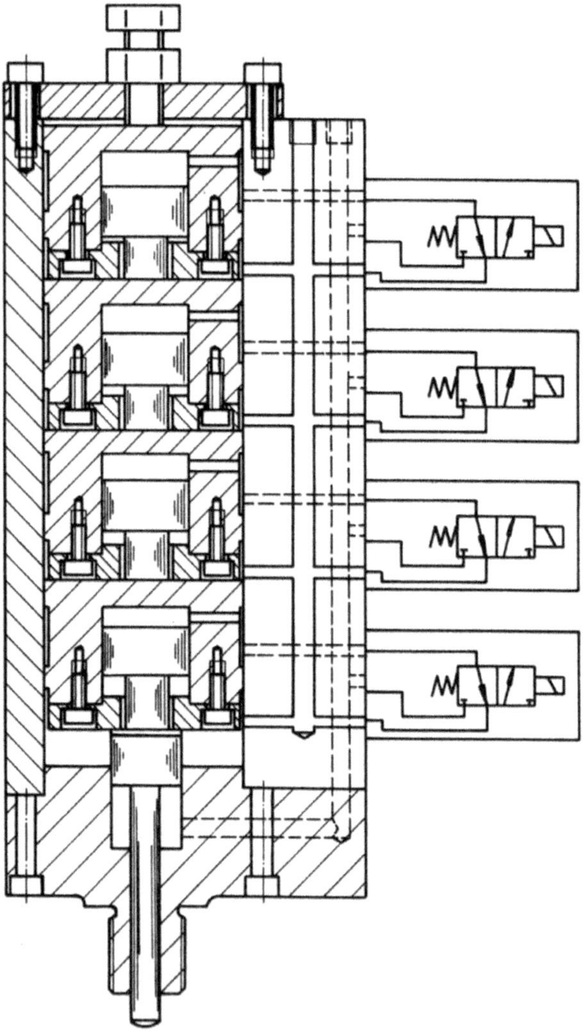

FIG. 13. Hydraulic digital to analogue converter.

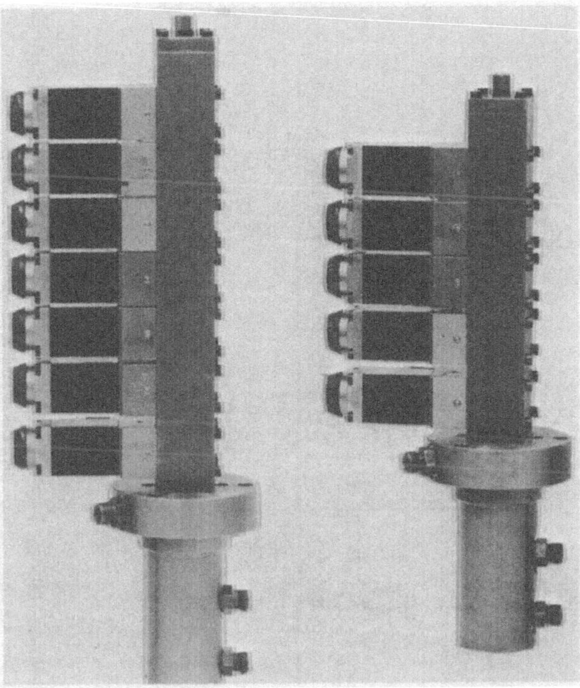

FIG. 14. Digital pressure valves for 31 and 127 pressure stages.

and voltage) is given by the stops in the final control cylinder, which are mechanically clearly defined. With a controlled oil temperature, and with a constant flow of control oil, measurements showed that changes in pressure did not exceed 0·5 % during 10^6 switching operations under normal operating conditions.

The constant maintenance of the pressure determined in this way in an upstream point, i.e. in the injection cylinder, is ensured by the functional principle of a precontrolled pressure valve which is a control element having a hydraulic return line within the equipment. Since, in the process of injection, the pressure control is only used in the hold-on pressure phase, i.e. in the static phase, the pressure in the injection cylinder corresponds exactly to the value limited or controlled at the digital valve.

4.2.2 *Digital Pressure Valve based on the Addition Principle*

It is also an obvious step to construct digital pressure valves without

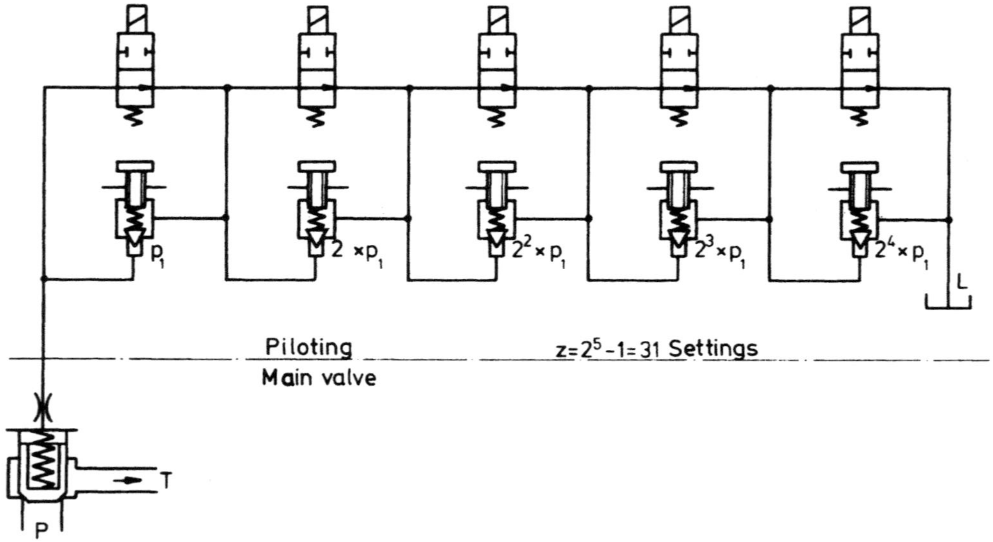

FIG. 15. System used to obtain digital pressure setting without using a digital-to-analogue converter.

resorting to additional energy and positioning cylinders. The solution shown in Fig. 15 adds up fixed pressures in any required combination. The individual fixed values are $p_1:p_2:p_3:p_4:p_5 = 1:2^1:2^2:2^3:2^4$. By combining the individual valves in various ways (see Fig. 10), 31 uniform pressure stages result when five valves are used. A digital pressure valve of this design can very easily be adjusted and adapted to suit various pressures in the system. With regard to reproducibility, this type of valve is similar to the valve in Section 4.2.1, given the same pre-conditions.

4.3 Digital Switching using Cartridges

In the injection moulding machine field, as well as in the industry concerned with the construction of presses, use is now being made of a control element that originated from early hydraulic concepts. For the water hydraulics engineer such a control element still represents the only possible way of blocking off low-viscosity pressure media in a reliable manner.[3,4] The control elements referred to are 2/2 conical seating valves or cartridges having the binary positions 'open' and 'shut' (2/2 means that there are two connections and two positions). The functions, which are rigidly linked in the conventional multi-way valve, can be individually controlled if they are broken down into controllable resistances of simple construction. With the aid of separate cartridges having a digital switching function—known as hydraulic 'bits'—reversing operations can be individually optimised.

4.3.1 *Control Problem Solutions*

Particularly difficult control problems can be solved very satisfactorily with cartridges. Here are a few examples:

(a) the supply and cutting-off of greatly differing volumes of oil in the supply and discharge connections of a consumer unit with unusually large surface area ratios such as are possible in the fully hydraulic clamp systems of injection moulding machines;
(b) controlled expansion of a volume of pressurised oil to avoid hydraulic shocks;
(c) protection of enclosed volumes of oil against undesirable increases in pressure caused by the action of external forces (e.g. forces due to inertia);
(d) creation of a counter-pressure (bias), independent of the flow in pipes, without the use of additional equipment.

The list could be continued. A particularly interesting aspect of these control elements is that it is now possible to use such units for pressure control so that one unit or system can control both path and pressure (see Section 4.3.3).

4.3.2 *Availability and Designation*

At the end of the 1960s, injection moulding machines having cartridges instead of pre-controlled multi-way spool valves were introduced for the first time in Germany with outstanding success. At that time use was made of the main piston of a pressure valve to construct the cartridge. Today there is a large number of well-known hydraulic equipment manufacturers

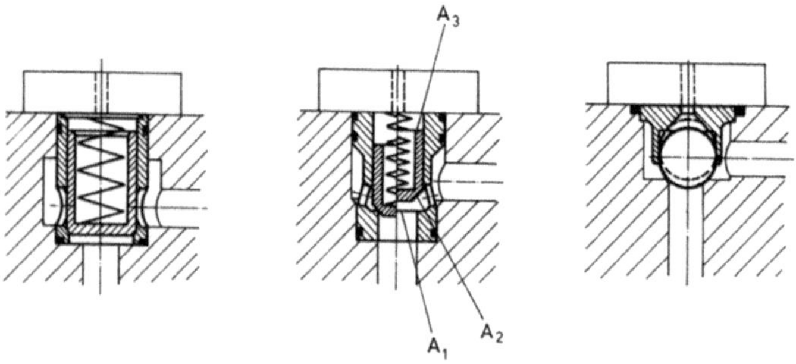

FIG. 16. Various designs of cartridge.

who produce such components and accessories in different designs (Fig. 16). Unfortunately, because of the rapid developments in this control technique, standardisation of dimensions and uniformity of designation have not yet taken place.

Instead of 'cartridges', one also speaks of 2/2 valves of seat or piston design, of logic elements, of logistors, or of individually controlled resistors. The very expression 'controlled resistor' gives the impression that hydraulic circuits formed with the aid of such components are in principle similar to transistor circuits in electronics, and thus implies uniformity within the broad art of control systems. This indicates why some conventional hydraulics engineers are critical of cartridge drawings which are now not so easy to understand. The absence of uniform or standard symbols has given free rein to fantasy and has led to a multiplicity of circuit-element symbols (Fig. 17). A description similar to that of a controlled non-return valve offers the best prospects for future standardisation.

4.3.3 *Design and Function of Cartridges*

The design and function of an insertable cartridge are illustrated in Fig.

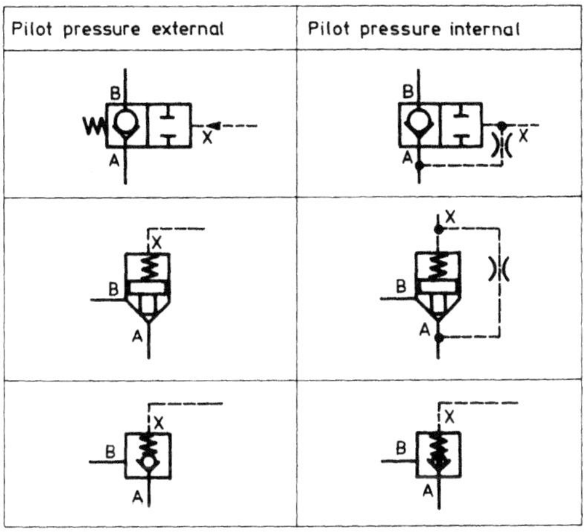

FIG. 17. Symbols used for cartridges.

18(a). A feature common to all the designs on the market is a valve seat which is actuated by means of a spring-loaded valve body. Flow resistance is dependent upon the control surfaces A_1, A_2 and A_3 and upon the pressures acting on these surfaces. The surfaces A_1 and A_2 are always acted upon by the pressure in the pipes A and B, whereas the control surface A_3 is acted upon by the control pressure p_3.[2] Therefore, in the cartridge, there are three different pressures (p_1, p_2 and p_3) and these act on the three surfaces A_1, A_2 and A_3 (see Fig. 18(a)). An additional force F_s, due to the spring, will also be considered, but other pressures or forces, e.g. those due to inertia, will be ignored at the present time.

The spring force is trying to close the valve. If oil under sufficient pressure is admitted to the top of the valve body then it will do so, since the forces acting in a downwards direction will be greater than those acting in an upwards direction. This may be shown by considering the following equation.

$$p_1A_1 + p_2A_2 - p_3A_3 - F_s = x$$

where x is the force on the operating element. Therefore when x is less than zero the valve is closed (see Fig. 18(b)). Where x is greater than zero then the valve will open. The complete stroke of the valve will be available for oil flow as soon as the control pressure p_3 has been switched off.

So far the cartridge element has been described as being fully open ($x > 0$) or fully closed ($x < 0$). It is now possible to use such elements in an

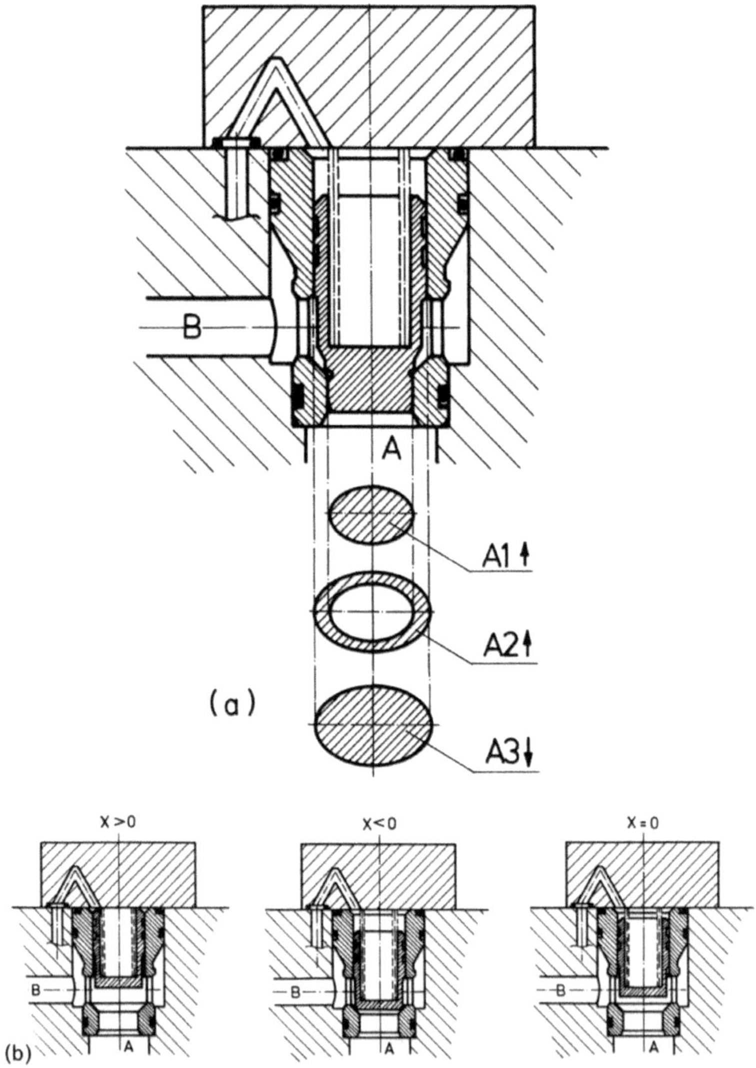

FIG. 18. (a) Control area of a 2/2 seating valve or cartridge. (b) Cartridge valve with valve in different positions.

intermediate position ($x = 0$) where the piston of the valve is in a position of floating equilibrium. By adjusting the pressures p_1 and/or p_3, this condition is disturbed and the piston changes its position. The resulting change in the cross-section between A and B (Fig. 18(a)) alters the flow resistance of the valve. This may be used to achieve different types of control function.

A pressure-limit or pressure-setting function may be obtained if a constant set pressure (p_3) is applied to surface A_3 by a pilot valve. If, for example, the outlet side of the valve (B in Fig. 18(a)) is connected to the tank then, if the pressure within the system exceeds the desired value, the pressure will be relieved.

Cartridges may also be used to achieve pressure reduction. In such a case the oil flows through the cartridge in the reverse direction, i.e. from B to A. The oil which leaves at position A is connected to control surface A_3 via a pilot valve which therefore reduces the pressure in A to the desired value.

For volume control the pressure difference can be controlled to a constant value by means of a calibration orifice. For this purpose the valve will be integrated in the controls as shown in Fig. 9.

It also has to be borne in mind when designing such a control system that in operation the pressures p_1, p_2 and p_3 can change and the required switching position can nevertheless still be obtained. The special feature of this new control technique is the possibility of providing a useful parallel and series connection and of combining the system with units for pre-controlling pressure. Thus it is possible to control path and pressure within one piece of equipment.

Depending upon the nature of the internal control oil feed, the surface ratio $A_1:A_2$ and the strength of the compression spring, different types of basic valve can be obtained from which all imaginable valve functions can be formed by wiring up the control connection in a logical manner. With the aid of an adjustable limitation of stroke, it is even possible to obtain throttle circuits or volume-control circuits. Exchangeable springs and control nozzles enable the switching conditions to be optimised even in difficult cases.

A previous exact analysis of the machine functions required enables control systems occupying a minimum of space to be designed with the help of the cartridge technique.

4.3.4 *Further Advantages of Cartridges*

Short switching times—because of the smaller parts of the switching element that have to be moved and because of short switching paths. The total switching time of these elements is approximately 20% of that of

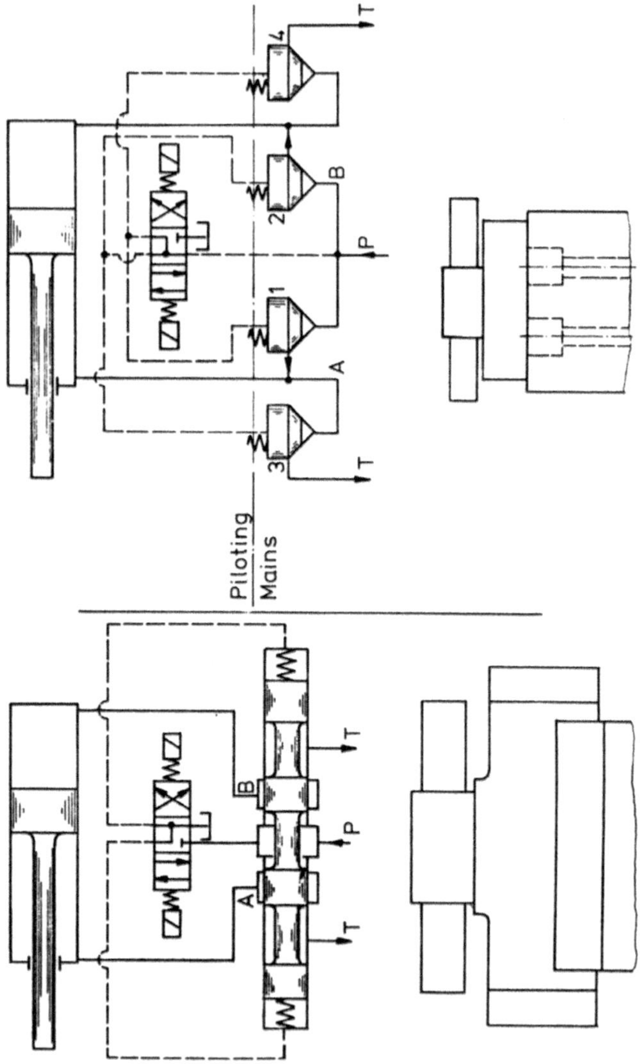

FIG. 19. Cylinder control by conventional and cartridge techniques.

comparable pre-controlled multi-way valves. Associated with this in high-speed machines are improved performance and greater precision in the operating cycle.

Gentler reversing operations—the cartridge opens immediately after the pre-control valve is switched without pressure peaks. In extreme cases use can be made of cartridges having shock-absorbing pins.

Reduced valve leakage—a control edge in the connection A (see Fig. 18(a)) is free from leaking oil: connection B has a seal which closes the gap between it and the control connection so that, compared with conventional multi-way valves, leakage is greatly reduced.

Low electrical control power—even in the case of high hydraulic power, only low electrical control power is needed, so electronic machine control can be used at low cost.

Reduced sensitivity to contamination—cartridge systems have reduced sensitivity to dirt and long service-life.

Low thermal losses—the large flow cross-sections result in low thermal losses.

The existing disadvantages are the non-standardised descriptions, the more complicated circuit drawings, and the fact that the controls are fully integrated into the hydraulic manifold, which means that location of faults is difficult. It is only a question of time before such problems will be solved, e.g. by standardisation and systematisation of the designs.

4.3.5 *Control Examples using 2/2 Cartridges*

Figure 19 illustrates the control system for a double-acting cylinder. The left-hand illustration shows the conventional spool control method, designed for a flow volume of approximately 150 litres/min. On the right the cartridge system for performing the same functions is shown for comparison purposes. The lower half of the figure shows a scale comparison of the sizes of the control blocks and equipment required.

An interesting example illustrating the switching arrangements that can be achieved with individually controlled cartridges is shown in Fig. 20. To obtain a similar multiplicity of flow values a large amount of valving, including many special elements, would be necessary in existing spool valve systems.

Three proposals using a hydraulic restriction (pilot) are shown in Fig. 21. The feature common to these circuits is the switching element which is almost completely hydrostatically balanced during the switching operation and is therefore independent of the actual operating pressure. This figure

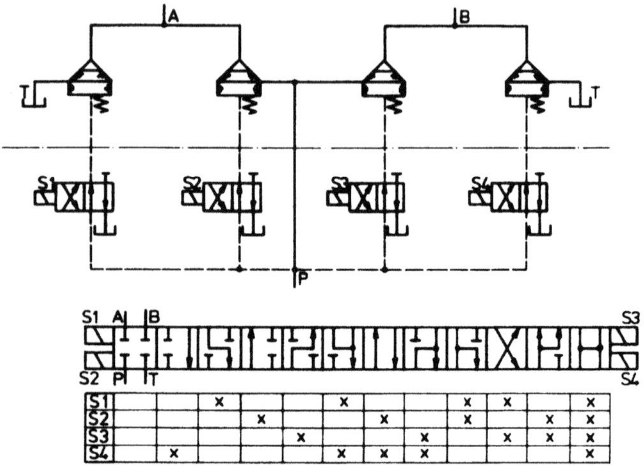

FIG. 20. Switching alternatives with four independently controlled cartridges.

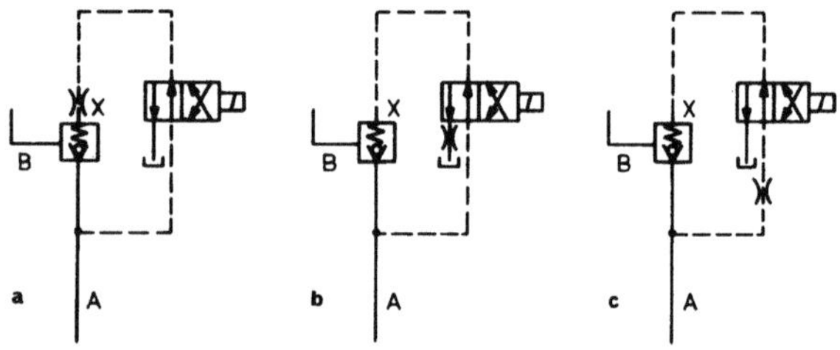

FIG. 21. Ways of influencing control times. (a) Opening and closing damped; (b) opening damped, closing not damped; (c) opening not damped, closing damped.

also shows how switching times can be varied depending upon the place where the control is fitted. To conclude this section on possible uses, reference is again made to the circuit diagram shown in Fig. 8.

5 PRACTICAL EXPERIENCE WITH DIGITALLY CONTROLLED INJECTION MOULDING MACHINES

Injection moulding machines with digital–hydraulic controls of the type described have been in successful practical use since 1975. Experiences to

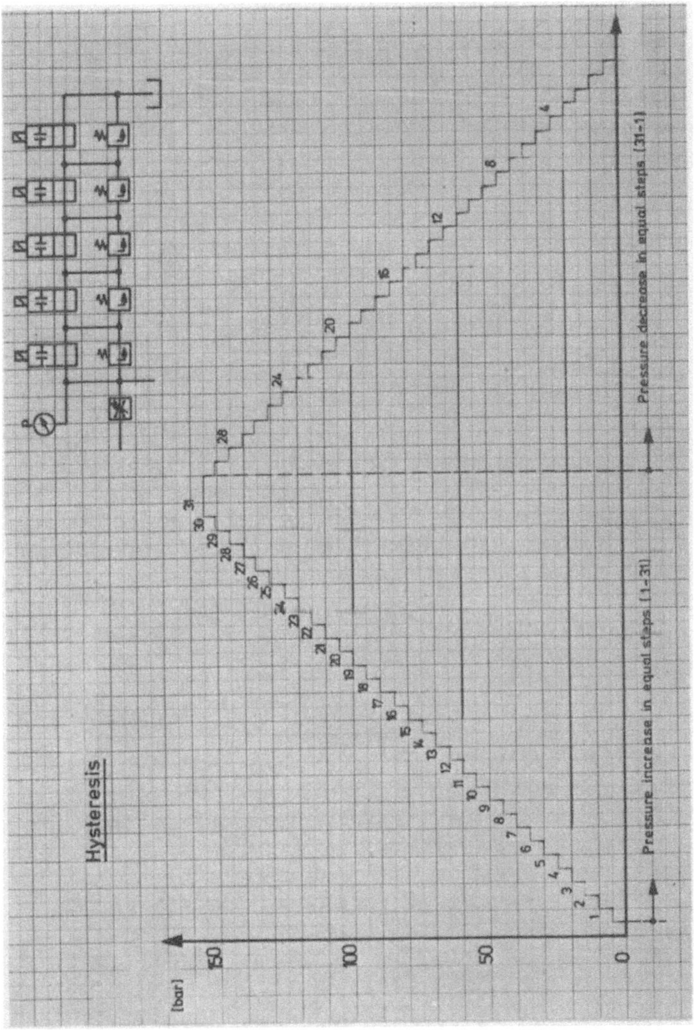

FIG. 22. Pressure stages and hysteresis of digital pressure valve used for injection and hold-on pressure.

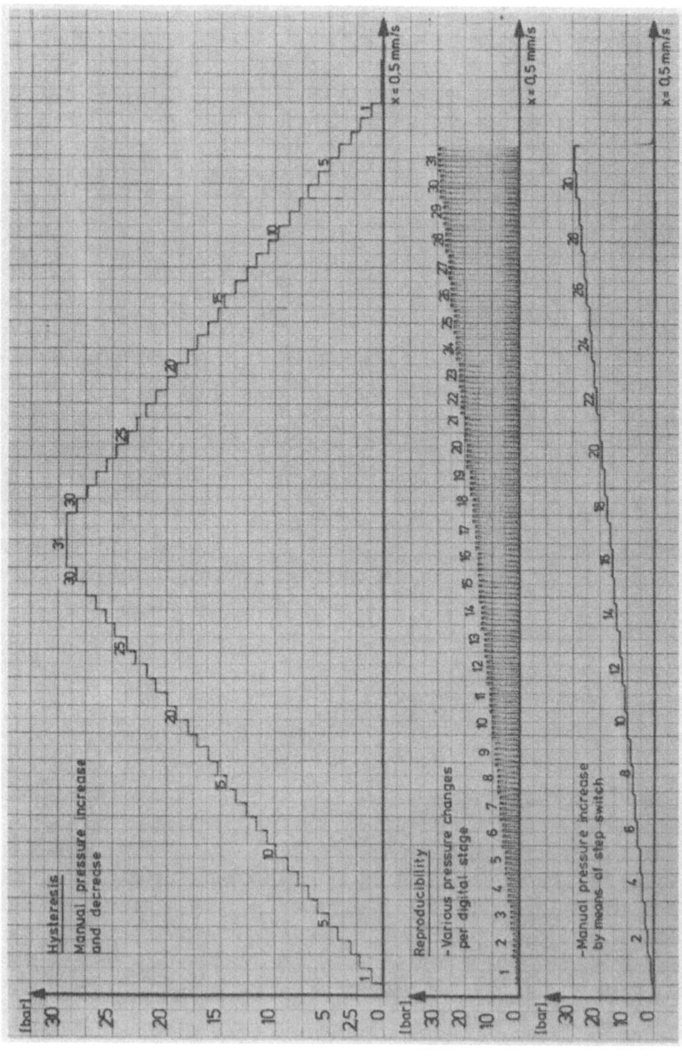

FIG. 23. Reproducibility and hysteresis of a finely stepped digital valve for the adjustment of back pressure.

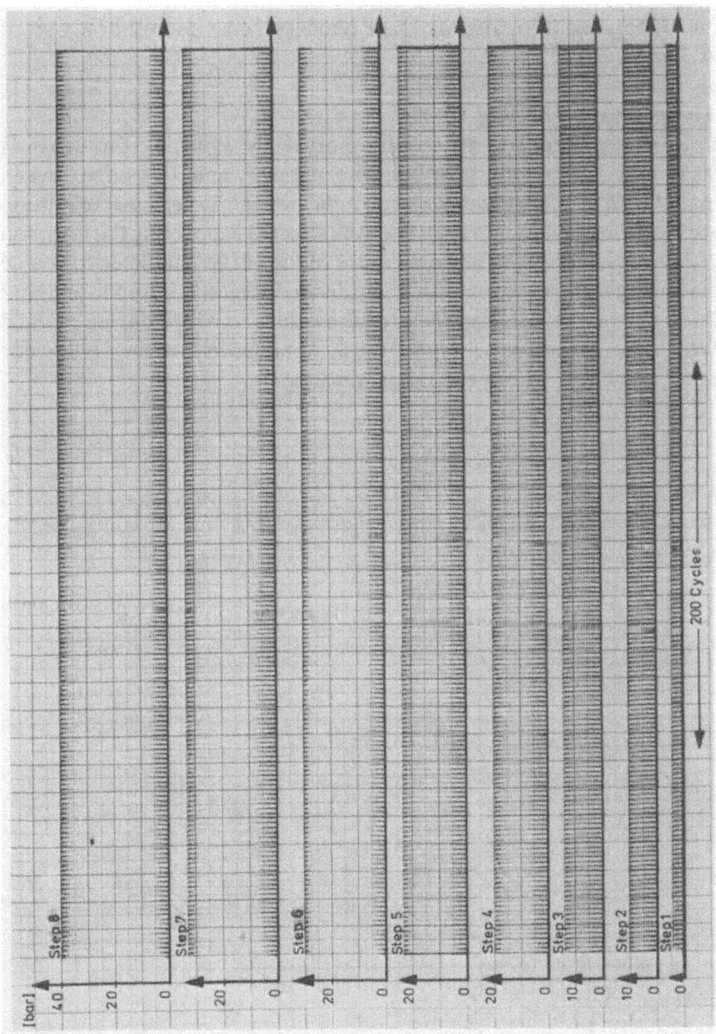

FIG. 24. Reproducibility of a digital pressure valve over the pressure step 1 to 8.

date are, on the whole, very encouraging. In particular, the results obtained on a production basis confirm those based on long-term tests, as regards reproducibility and freedom from hysteresis in controlling all pressure and volume parameters. To illustrate these remarks, some test results are given below.

5.1 Hysteresis-free Setting

A pressure characteristic curve was plotted by adjusting the controls upwards and downwards in a stepwise manner, and this curve, when compared with similar pressure stages, indicated the hysteresis-free setting. Figure 22 shows a pressure range that was selected as covering the injection pressure/hold-on pressure setting. Figure 23 shows this characteristic curve for the back-pressure setting with finely divided stages. It is also found that no difference occurs between the pressure values recorded during manual stepwise upward adjustment and those obtained by setting a specific pressure, anywhere within the range, directly.

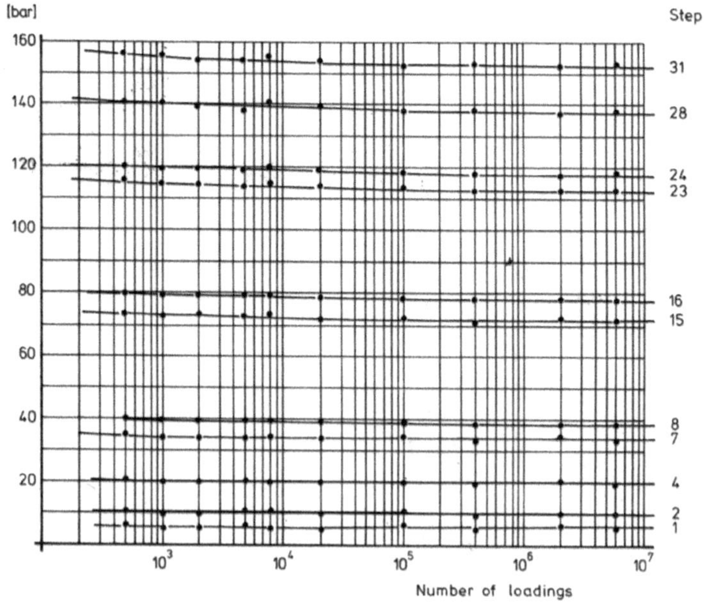

FIG. 25. Long-term test result on a digital pressure valve.

5.2 Reproducibility and Long-term Repeatability

A requirement associated with a high-duty injection moulding machine is that the set parameters should always have the same values from one shot to another and over fairly lengthy periods. Figure 24 illustrates an automatic recording of several different pressure settings which were repeatedly obtained over a fairly lengthy period by adjustment from zero. No measurable tolerances could be detected. Long-term tests showed, for a load cycle of approximately 6×10^6, only a slight drop (approximately 1·5 %) in the predetermined pressures. This change occurs during the first load cycle and is attributable to the settling phenomena in the mechanical components of the equipment (Fig. 25).

5.3 Reduction of Setting Times and Weight Tolerances

The coded signals for the hydraulic machine parameters enable optimum values, once established, to be reset in a reliable manner. Thus, for example, it can be assumed that any differences in the manometer readings are usually caused by the manometer itself, which, as is well known, undergoes wear. Similarly, vernier settings on volume regulators or potentiometers are affected by reading and setting errors, and are inferior to the digital orifice-control system. The comparison illustrated in Fig. 26 shows the time saved when restarting production with digital machines.

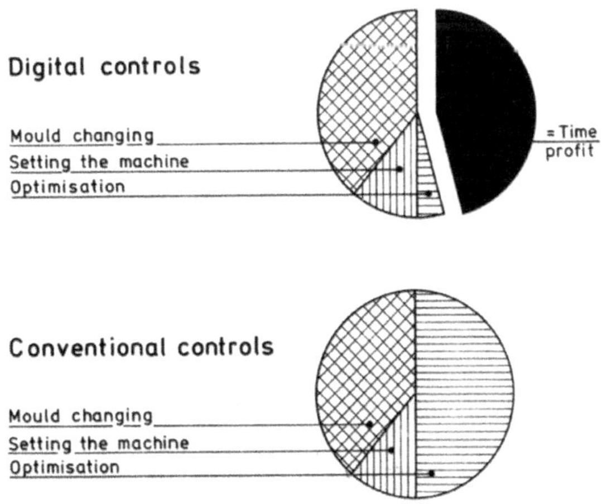

FIG. 26. Decrease of starting time for digitally controlled injection moulding machines.

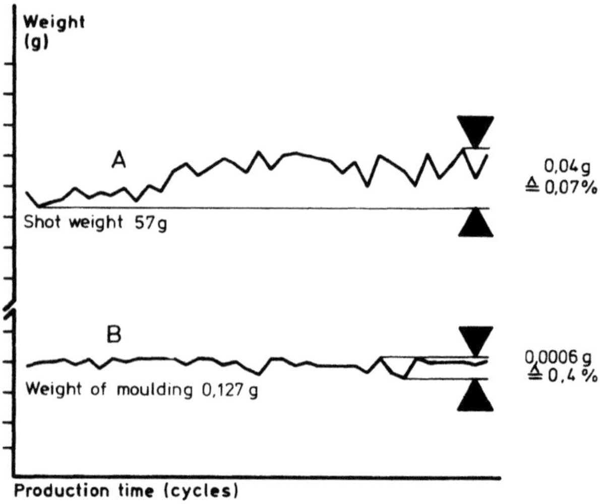

FIG. 27. Weight tolerances in production. A obtained from a four-impression tool; B obtained from a six-impression tool.

In addition to short start-up times, these machines provide finished parts of high quality. Two examples (Fig. 27) from the production of optical parts show the attainable weight tolerances which should be assessed in conjunction with the weight of the part.

6 CONCLUSIONS

The requirements imposed on modern injection moulding machines today are, apart from technical data required by the market, a high degree of automatisation, more machine functions, and accurate reproducibility of the machine parameters under operating conditions.

The use of the digital technique in the hydraulic control system has led to the development of digitally adjustable pressure and speed valves as well as to the use of 2/2 insertable cartridges for controlling the movements of the machine. Injection moulding machines in which this technique has been used are distinguished in particular by accurate reproducibility, smooth running and ease of supervision and maintenance.

REFERENCES

1. *Hydraulik für Kunststoffverarbeitungs-Maschinen*, Conference, Lohr/Main 1977, pp. 14–25 and pp. 43–7.
2. Lange, H. J. (1976). Spritzgiessen—Steuern oder Regeln? *Kunststoffberater*, **6**, 260–6.
3. Blüml, H. (1975). Hydrostatik bei kunststoffverarbeitenden Maschinen, *VDI-Berichte* No. 228, 125–33.
4. Cartridge-Elemente und -Steuerungen, *Fluids*, September 1975, 34–8.

Chapter 2

CLAMPING SYSTEMS

R. A. IRELAND

Technical Director, Ketch Plastics Ltd, Lichfield, Staffs., UK

SUMMARY

The most common clamping systems are considered in detail, both in comparison with each other and as they affect the normal process of injection moulding. Mechanical, hydraulic and control functions are discussed as is the effect of modern technology on these functions.

Because of the very large variations in the basic systems it has not been possible to discuss individual machines, and many manufacturers produce machinery involving combinations of techniques not considered in this chapter. However, the basic principles are common to all machines and it is the understanding of these principles that the text attempts to provide.

Short sections on some of the more unusual locking systems are included in order that the reader may appreciate the lengths that designers have gone to in order to fulfil special requirements.

It has not been the aim to cover all known systems, either past or present, and the author apologises to any designer whose methods have been excluded.

To the reader the author would wish to convey the thought that all systems have both advantages and disadvantages, which must always be considered in conjunction with the type of work they are to be asked to perform.

1 INTRODUCTION

Unlike the single reciprocating screw injection unit, which, after many years of development, has become almost universally accepted as being the most satisfactory compromise, the clamping end of the modern injection machine still has almost as many variations as ever.

There are many reasons for this state of affairs and it is unlikely that the next decade will see much alteration. Probably the most important aspect is that of manufacturing capability; a company set up to machine toggles accurately is unlikely to have the boring and grinding equipment available to enable them to cope with the manufacture of a large cylinder assembly for a direct hydraulic lock machine. It is therefore to be expected that such a company will enter the large machine market with a lock and block clamping system. Conversely a company making a large direct hydraulic lock may well compete effectively against toggles in the small machine end of the market, using their basic design scaled down.

As the industry stands at this moment there is no one correct answer to the question, 'What sort of lock should I buy?'. The type of moulding that is to be manufactured with particular emphasis on the depth of draw will have some bearing on the choice, but the decision on press manufacturer will more often than not decide the type of clamping system finally selected.

There are some exceptions that stand out, primarily due to specific developments, and the resultant machines have been universally accepted as leaders in their fields. For instance, the trend towards thin-walled disposable articles resulted in the very high speed toggle machines with open–close–open speeds of around 1–2 s. Whereas this speed can be approached by the direct hydraulic clamp on short strokes, on the movement necessary to clear, for instance, the standard vending cup, the direct clamp falls short on speed, primarily as a result of the volume of oil involved in the clamping cylinder. Perhaps the most interesting development that is influencing the clamp design is the fairly recent trend towards thermosetting injection. The necessity to supply machines with 'breath' facilities has resulted in some interesting control problems and although these have been overcome by various means by the manufacturers of toggle machines, the direct hydraulic machine makers have been able to offer this facility very easily. The 'lock and block' machine already had the necessary action, requiring only additional timers to control the operations.

The purist will undoubtedly object to the term 'lock and block' as used in this text and in essence he will be correct, but it has become a phrase so universally accepted as describing any form of two-stage locking system that the author has no hesitation in using it to describe such machines in general terms. There have been many attempts, some more successful than others, at manufacturing different types of locking systems, and although this chapter deals primarily with the conventionally accepted system, a brief description of some interesting developments is included at the end, which may be of interest to readers faced with problems not catered for by more conventional equipment.

2 CLAMP UNIT REQUIREMENTS

Whatever movement system is used, every clamp unit needs to conform to certain requirements and it is necessary to study these in detail in order to relate the advantages and disadvantages of the different types to the basic criteria. The basic requirements for every unit are:

(i) they must be safely guarded with adequate interlocking;
(ii) both platens are to remain parallel throughout the whole movement;
(iii) there is ample locking force, preferably no less than 7 tonnes per square inch of projected moulding area;
(iv) platens are to be of proportions such that minimum deflection occurs at full tonnage with small or offset tools fitted;
(v) there are ample tool fixing facilities (minimum to Euromap standards);
(vi) there are ample ejection points—centre ejection only is not satisfactory;
(vii) movement speeds are adjustable and transition between speeds is smooth;
(viii) there is easy and repeatable stroke and die height adjustment; and
(ix) easy removal of at least one tie bar.

Additional requirements, depending on type of tooling to be used, are that there are:

(x) one or more intermediate halt position for core pulling;
(xi) die halt in part-closed position for insert loading;
(xii) hydraulic ejection equipment; and
(xiii) facilities for mould breathing.

Some further explanation of certain of these points is necessary and these will be discussed in Sections 2.1 and 2.2.

2.1 Basic Requirements

2.1.1 *Safety Considerations*

New legislation in the UK has caused an almost total redesign of safety systems and it is only the latest types that are considered suitable for discussion here.

All types of press should have three totally independent safety systems and it is now almost universally accepted that these should be electrical, hydraulic and mechanical. It is important to note that any system must arrest the press movement, at any point during the stroke, independently of either of the other systems. This precludes the use of a fixed stop bar, as this is inoperative once the press has started to close. Therefore the fitting of a so-called progressive scotch is a requirement in any modern press. Because of difficulties in manufacture and the lack, as yet, of universal acceptance of the need for a progressive scotch, few presses are equipped with such safety devices. It is only a matter of time, however, before all new equipment sold in the UK conforms to this additional safety precaution.

Electrical safety is usually adequately covered by the use of limit switches. Two are essential and these must be fitted so that they operate in dissimilar modes and are connected to circuitry that detects malfunction of either switch.

The situation is not quite so simple when considering hydraulic safety. Much depends on the basic circuitry, but the common method, until recently, of returning the pilot valve oil to tank, in the guard-open condition, is definitely not satisfactory. Both blocking the main oil line to the closing cylinder and opening the same circuit to the tank are used and give adequate safety. A combination of both is probably the most correct solution.

All three separate safety systems must be operated by very small guard movements. In order to ensure reliable operation spring movement in either direction should not be involved. To ensure safety in the event of mechanical failure of the guard mechanism, the hydraulic safety valve should be pressure-loaded to the safe condition.

The interlocking of safety screens, other than the main guard, must be equally effective. However, due to more infrequent operation such interlocking can be less elaborate. If the back guard is to be interlocked with the main guard, this must be mechanically strong enough to ensure that no malfunction can occur, even under the strains imposed by an energetic tool setter.

In some machines it is necessary to move the front guard behind the front face of the moving platen in order to set ejector switches, etc; in this case it must be arranged so that this facility is made mechanically impossible during normal working.

The argument as to whether the press should open or stop when the guard is moved will never be satisfactorily resolved. If mouldings have to be removed by hand from the tool it becomes almost essential to have a press that can continue to open with the guard in the open condition. If a part-closed die halt position is required for insert loading a special guard interlock may be essential.

The guard is, perhaps, the most abused mechanism on any press and because of this and its legal implications, the total system should be inspected on a daily basis.

2.1.2 *Parallel Platens*

In order to ensure that the platens remain parallel at all times, the moving platen bushes must be as long as possible. Accurate location of both the tie rod nuts and the pressure collars must be ensured and very special attention must be paid to the correct refitting of these components, after a tie rod removal. Machine manufacturers in most cases put lining-up marks on all components to enable easy refitting, and it is essential that these are correctly used at all times.

Although not by any means universally accepted, it is more likely that a press will remain 'square' if fitted with platen support slides. They take the full weight of the platen and the moving half of the tool and leave the tie rods to perform their two main functions of taking the strain and guiding the platen.

All moving parts must be capable of being lubricated automatically and if support slides are used these must be adjustable for wear.

2.1.3 *Locking Force*

It goes without saying that a 400-tonne machine must develop a locking force of at least 400 tonnes, but what is not so readily accepted is that, with today's high-performance engineering polymers, an effective closing force of 6–7 tonnes/in^2 (90–110 MN/m^2) is often necessary to ensure high-quality flash-free mouldings.

The average toolmaker will complain bitterly to the moulder who crushes his flash line. It is seldom accepted that perhaps this could have been avoided by more locking force, as it is a very common occurrence for a 'rolled edge' to appear after very little use and this is usually the effect of minute flashing. The effect is progressive and, of course, the loads applied to the flash line increase significantly as the progression takes place—so much so that the tool may have to be fitted to a larger machine to 'shut it out' again. The toolmaker by this time has given up all hope of educating the moulder and charges double for repairing the tool. The fault lies not only in the lack of co-operation in the beginning, but also in the lack of appreciation on the moulder's part of the importance of having enough locking force to stop *all* flash.

The rule-of-thumb method of multiplying the projected area by injection pressure to arrive at a locking force, is all right in theory, but has proved inadequate in practice so often that a safety factor must be added to the equation. To start with, the 'injection pressure' must be the maximum the press can deliver, not the one thought necessary to make the moulding in question. Invariably during production the melt flow index (MFI) of the material being moulded will alter and more pressure than originally envisaged will be used to the detriment of the tool. Secondly the depth of the moulding is not catered for in the formula, for although the tooling engineer may well have based his expected injection pressure requirements on the flow length to thickness ratio of the moulding in question, the result is still likely to be on the low side, because of frictional forces on the material flow (few mouldings have ideal shapes) and the fact that little if anything will be known about the viscosity of the melt at different parts of the tool until the first few mouldings have been produced.

It may seem to the uninitiated reader that there are no rules. This is not so; rather, the rules are dictated by so many factors, both in relation to material and tool design, that there cannot be one rule only but a whole series, coupled with a good deal of experience. Perhaps the most important point is never to be tempted to use a press that has insufficient locking force. At least 15% should be added to the calculations, thus giving the moulding engineer a chance to use the injection end controls effectively so that finished product sizes and shapes can be achieved and maintained in production.

Ideally the tool only has to be closed so that the flash line is in intimate contact. No force needs to be applied if the toolmaker has done his job correctly. So long as the tool is now physically restrained from moving no flash will occur. Unfortunately Utopia is not yet an engineering term: tool steel does compress and tie bars stretch, so much so that tool faces in practice have to withstand the full locking force of the machine. The force on the flash face gradually diminishes as the tool is filled, and reaches a minimum at the point of high pressure injection cut-off. If, at this point, the locking pressure on the tool face is insufficient, the tool will flash and it will be necessary to apply more locking force or reduce the area of the flash face. However, a practical balance must be achieved if tool damage is to be avoided and it is here that the closest co-operation between the moulder and the toolmaker is essential.

2.1.4 *Platen Thickness and Mould Mounting*

The thickness of the platens is of paramount importance to ensure that deflections caused by any unequal loadings are kept to a minimum. There is a strong temptation to reduce thickness because of the ever increasing search for more speed and, in consequence, reduction in reciprocating weight. It is also important that the aperture into which the injection barrel fits is not over large. This has proved to be a weak area in many machines in the past.

With reference to the method of tool fixing, it must be remembered that the total strength of the platen is reduced with every hole and a good balance between the number of holes and the weaknesses that they cause must be arrived at in order to satisfy the performance and the versatility of the machine. The total effect is complicated by the necessity to supply ample ejection facilities on the moving platen, which is inherently most likely to fail and is also the one reduced in thickness to the minimum in the interests of speed.

Two points often overlooked are that a 'T' slotted platen requires to be about 20% thicker than a drilled one for the same stiffness and that every threaded hole is a multiple stress raiser. In consequence, any hole should only be as deep as is dictated by safety.

2.1.5 *Clamp Stroke Setting*

Too many good machines have been spoiled from the user's point of view by poor layout of controls. This is most obvious in the locking area, where, in some cases, settings can only be satisfactorily carried out by two persons. The result of using modern electronic control has been two-fold, firstly in bringing about the availability of fast digital setting of all press functions and secondly in forcing press manufacturers, even those using more mundane systems, to tidy up the controls making them more ergonomically acceptable.

It cannot be over stressed that the correct positioning and the consistent repeatability of all platen movements is essential if a well designed locking unit is to get shop-floor acceptance.

The latest equipment giving electronic control of speeds, pressures and, more recently, movements, is to be highly recommended and the availability of memory circuitry gives guaranteed repeatability. The trend towards fitting all these setting controls into a remote cabinet needs to be carefully considered, however, as it may prove impossible to obtain initial settings without assistance. This will be most critical on the medium high speed machines, the very machines where the most advantage can be gained by using this type of control, but also the one where, because of lack of space and vibration, the designer is forced to fit remote cabinets.

2.1.6 *Tie Bar Removal*

The ease with which a tie bar, or preferably two, can be removed, is of paramount importance in some moulding shops whereas in others it is of little or no consequence. The method incorporated to remove and re-fit tie rods depends upon the type of lock unit involved and will be discussed under these headings, but certain criteria are common to all types. The points which should be considered if it is expected to habitually remove a tie rod are:

(a) Will there be a likelihood of thread damage, with the bar in the 'out' position, especially if this means that the bar is left hanging in the fixed platen?
(b) Can the distance collars be easily removed without stripping other equipment from the machine?
(c) Are the nuts easily accessible both to turn by hand and to tighten finally by the tool provided?
(d) Is a tool provided?
(e) If the nuts are too large for easy handling are they fitted with suitable lifting points and can they be easily replaced on the threads? Some form of relocation device is essential in such cases.
(f) Can the bar be accurately retightened to the correct preload?

Tie bar breakage today is rarely attributable to incorrect design, but is more likely to occur because of incorrect preloading. The designer cannot, however, be completely exonerated if it is difficult to remove and accurately re-fit a tie rod to his machine.

2.2 Additional Requirements

These are more important to the specialist moulder. However, it is becoming more common to have at least one position of core pull as a standard on most of the bigger machines. The only point that should be looked at carefully for both the core pull and die halt is that the movement to the chosen position is smooth. Facilities for mould breathing are discussed in later sections. The author feels that hydraulic ejection should be standard to every press and it should be fitted so that a link-up with the ejector frame of the tool is possible. The positions should be multiple rather than a single one just in the centre. If hydraulic ejection is fitted it is almost universally accepted that this should now have more than one stroke. It is important that its stroke can be adjusted. An added advantage is to have a pressure-reducing valve in the circuitry so that damage to the tool's ejection system can be avoided if a malfunction should occur.

3 DIRECT HYDRAULIC LOCKING SYSTEMS

As the name implies, locking pressure is applied directly from a hydraulic cylinder with no other mechanical mechanisms. Simplicity is perhaps the outstanding advantage of this system, but against this must be set the problems associated with moving large volumes of oil in and out of the main cylinder. All control is concerned with the volume and pressure of the oil and so it is in this area that complications occur.

3.1 Mechanical Considerations

3.1.1 *Piston Alignment*

Looking at the mechanical side first, it will be readily understood that moving with the platen will be the not inconsiderable mass of a hydraulic piston. Part of this mass has to be borne by the moving platen when it is in the closed position, the rest—and an ever increasing percentage—by the gland guide, at the front of the cylinder, as the press opens.

Under ideal conditions the gland guide should always support the total weight of the piston, but unfortunately in practice this has rarely proved possible. This has given rise to problems in alignment between the front face of the piston and the rear face of the platen. This area is usually increased in complexity by the necessity of providing room for the ejection mechanism. A compromise has to be made which usually results in a very constricted ejector assembly. There are, however, exceptions and good examples of well supported pistons are available from some manufacturers, the support in the gland area being the critical point.

3.1.2 *Large Machines*

The piston has of course to be enclosed in a cylinder and this in itself will be a fairly massive part of the machine, its construction often being the limiting factor of a manufacturer's press range. In order to keep the sheer size of this area within practical bounds, some manufacturers have resorted to more than one cylinder and this has the added advantage of giving more room at the back of the platen, but at the same time necessitating more careful stressing of the platen due to the higher localised loading.

On machines fitted with large pistons, replacement of piston rings or seals may prove to be difficult. The only practical solution of assembly is to perform this operation in the vertical direction, necessitating the complete removal of the assembly from the machine frame. On smaller sizes this consideration is inconsequential, but care must always be taken to ensure complete alignment of all parts if reliable operation is to be expected.

In the past it was common to fit the locking cylinder with a straight displacement ram, reverse motion being obtained by supplementary cylinders. This method reduced the importance of good alignment, removed the assembly problems, and allowed larger machines to be built on existing equipment, due to the reduced requirement of boring. However, it is rare to see such equipment today; the author knows of no such press on the market from a European manufacturer.

3.1.3 *Tie Bar Stretching*

On the straight hydraulic locking unit the tie bar should have no other work to do than that of guiding the platen and of taking the locking stresses. If correctly aligned, no side forces, other than the weight of the platen, should be exerted on the bars and they are not required to stretch to supply locking force. However, they will, in fact, stretch, and it is still necessary to allow one or other of the fixed ends to move in relation to the press bed. Due to the previously mentioned mass of the hydraulic cylinder it is more normal to allow the fixed platen of this type of press to move (in contrast to that in a toggle press which will be discussed later). With the above facts in mind some manufacturers make their tie rods much larger in diameter than would be the case on a toggle machine of similar size and, by using longer guide bushes, rely on these to support the weight of the platen/tool assembly. The effectiveness of this method is very dependent upon the size of the machine, but is not entirely satisfactory as no wear adjustment can be accommodated. It is always better to have a machine with platen support guides.

3.1.4 *Stroke and Mould Height Problems*

As can readily be seen, a conventional direct hydraulic lock has the inherent characteristic of decreasing maximum stroke as tool closed height increases. To overcome this problem press designers often produce equipment with longer maximum strokes than equivalent toggle machines in order to compete in the market place. Various methods of giving differing mould heights are offered. The commonest uses spacers between the piston and the moving platen. Unfortunately this area is the one used for ejection and it is often not possible to obtain maximum ejection stroke with maximum mould height. This area is perhaps the most unsatisfactory, with this type of machine, from the moulder's point of view.

Another method of overcoming the stroke problem has been attempted by at least one manufacturer and although very satisfactory, it creates the added complication of having to allow the whole locking assembly to slide on the machine bed. This method employs collars, which can be fitted around the tie bars, and these can be moved either behind or in front of the stationary platen, so making the tie bars effectively longer or shorter.

3.1.5 *Summary of Mechanical Considerations*

In conclusion, on the mechanical side it should be obvious that the direct hydraulic lock system is very simple, applies no unwanted forces to the press, and, if correctly made, is reliable. However, stroke adjustment is difficult for a large range of tool heights. This could be of importance to a moulder who has a very wide range of tooling, although there are always some restrictions in both the stroke and distribution of ejector systems.

Tie bar removal and re-fitting in this type of press very rarely cause any difficulties because of the fixed nature of both ends of the tie rods.

3.2 Direct Clamp Hydraulics

If no criteria other than opening and closing at constant slow speed are required the hydraulic circuitry necessary on this type of locking unit is the ultimate in simplicity and on very small machines simple circuitry has been successfully used.

3.2.1 *Variable Speed Locking Systems*

The requirement for more speed and variations in this speed has, however, created more complications. This is particularly true for mould closing where it is normally necessary to slow down the movement as the mould faces approach one another. Constantly variable speeds over an infinite range can be obtained by the use of variable output pumps. However, in the majority of machines, movement is controlled in steps by flow control valves. Unfortunately these are seldom shock-free so that on the higher speed presses the platen moves in a series of jerks.

As the press size gets larger the jerks are smoothed to some degree by the inertia of the moving parts. It is only over the last two or three years, with the advent of the electronically operated infinitely variable control valve, that a completely smooth hydraulic lock system has become feasible. It is, of course, necessary to control this type of system by means of position transducers with associated electronic equipment. This system allows ideal variable speeds to be obtained and, on the smaller sizes of hydraulic lock machines, a higher dry cycle speed.

3.2.2 *Prefill Systems*

Theoretically larger machines can be made to work in the same manner as the simplest form; however, in the interests of economy, this is never attempted. A method of filling these large cylinders with oil, other than directly from the pump, is necessary. This oil is now universally obtained via a so-called pre-fill valve which connects the cylinder to a volume of oil at atmospheric pressure. It has become customary, but by no means universal, to contain this oil in a tank situated over the locking cylinder, thus obtaining 'always flooded' conditions to minimise air dilution.

The circuitry has now become very different. In addition to the main cylinder an extra one of much smaller dimensions is now necessary to give movement. However, the main annulus is still available, being small in volume, for the opening stroke. The small cylinder, or cylinders, have to be supplied with oil at high pressure to enable high acceleration rates to be achieved. Although their volumes are relatively small, control of speed is just as difficult as in the simple form of press but mould protection can become very easy due to the relatively low total pressure exerted by the small cylinders. Variations on the basic concept have been used to enhance the acceleration and give better control, and most machines are not quite as simple as this text indicates. Each manufacturer has his own modifications, most of which give significant improvements over the operation of the basic machine.

Regardless of what modifications are fitted to give better control and faster speeds, it is still necessary to close the pre fill valve and pressurise the main cylinder. This takes time and it is very surprising how much oil has to be pumped into a big cylinder to give full locking pressure. It is normal to use pressures of up to 400 bar (2·6 tonnes/in^2) in the hydraulic cylinder. To get these pressures without resorting to high pressure pumps some form of intensifier is required. This increases the speed problem. Instead of the single function of the simple lock, the modern lock unit goes through at least five functions: energise accelerator rams, decelerate rams, close pre-fill, apply pump pressure, apply intensified pressure; this does not include any speed variations on the forward stroke.

Opening, however, is relatively simple and only the large mass of moving parts gives any problems to the designer. In most cases valves used on the closing stroke are also used for control of the opening movement.

3.3 Direct Lock Characteristics

To sum up, the following main points must be considered on a direct hydraulic lock machine:

1. simple mechanical layout;
2. lubrication restricted to platen bushes only;
3. large mass of moving parts;
4. no unwanted forces on tie bars;
5. no unwanted forces on platen;
6. large pressure area on platen;
7. restriction of ejector assembly;
8. restriction of mould height stroke adjustment;
9. complication of hydraulic circuitry to obtain satisfactory speed control;
10. lack of smoothness at high speed except on the latest presses;
11. difficult maintenance in the unlikely event of serious breakdowns on larger machines.

4 TOGGLE LOCK MACHINES

Unlike the direct hydraulic lock machines the complication for this type is on the mechanical rather than on the hydraulic side. That is not to say that control is lacking in the modern machine, but rather, because of the inherent movement created by a toggle system, it is usually unnecessary to fit anything but simple flow control valves to give the basic speeds that are required.

4.1 Toggle Lock Hydraulics

The volume of oil required to operate the locking ram is very small when compared to that involved in the direct hydraulic machine and invariably comes directly from the pump. Independent control of opening and closing speeds is desirable and easily fitted. It is sometimes thought necessary to further restrict the opening speed over the final part of the movement if mechanical ejection systems are to be used. Cushioning of the locking cylinder, in the mould-closed direction, is adequate to avoid unnecessary shocks. It is common practice to reduce the stroke of such a machine by means of a screw-in stroke limiter on this main cylinder, this device invariably carrying the damper assembly to cushion the final opening. It is usual to fit some form of reduced pressure valve to operate over an adjustable distance, on the final closing stroke, to work as a mould protection device.

4.1.1 *Die Halt Positioning*

The problems only start to arise on the hydraulic system when die halt positions are required and it is in this area that the press designer will have most difficulty in getting satisfactory smooth working, without resorting to restricted speeds. It will readily be seen that any toggle system gives a desirable acceleration/deceleration curve in both directions, but at the mid-point when the platen is moving at the greatest speed the shock loads applied in stopping the motion must be smoothed out. This must be done when the mechanical advantage, and therefore movement, of the operating cylinder is at its lowest. Any die halt positions necessary to allow for core pulling or other operations should be designed to occur as far away from the mid-stroke point of a toggle machine as possible. Again, like the direct hydraulic machine, the situation can be alleviated to a great extent by making use of the advantages gained with electro-mechanical flow control valves coupled to an electronic positioning control. In any event, if a machine is to be used to any great extent with a die halt position it is essential for long trouble-free life to ensure that the motions are smooth. In the majority of cases it may be desirable to reduce the speeds in order to obtain the required motion. It is far better, if perhaps a little more complicated, to arrange core pulling to operate between two points on the stroke, the speed of platen movement being adjusted so that core pull can occur with a satisfactory safety factor between the two set-points. In this event the movement will be continuous and smooth, any jerks being confined to machine or tool malfunction. In this event it is, of course, essential that the press will stop in time to avoid damage if a core malfunction should occur.

4.1.2 *Control Layout*

The general advantages of modern control systems can be applied to a toggle machine and the remarks on ergonomic layout are equally true for this type. Positional controls may be somewhat easier than on a direct hydraulic machine since magnified motion is obtained from the locking cylinder, and, as long as it is remembered that this motion is dissimilar to the actual platen motion, accurate positional control is available.

4.2 Mechanical Considerations

Whereas the hydraulic locking machine had very few moving mechanical parts, the toggle machine has many, and in general it is in this area where the skill of the designer is important in order to get longevity. The two most common basic layouts are shown in Figs. 1 and 2. Terminology varies widely between machine manufacturers and in order to simplify the text the various major parts of the toggle have been shown in Figs. 1 and 2.

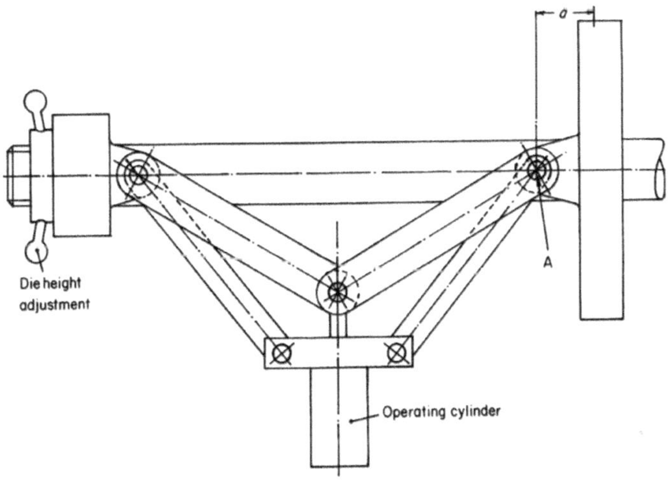

FIG. 1 Single toggle system.

4.2.1 *Single Toggle Systems*

Dealing with the single mechanism first, it will readily be seen that there are some unwanted forces acting on the moving platen and unless the support for this platen is carefully designed, considerable wear will take place in a short working period. The forces acting on the front link pin A vary in direction as the lock closes and there is a resultant force trying to twist the platen around the guide bushes. It is therefore necessary to have the guide bushes as long as possible in order to reduce the distance *a*. In most present-day single toggle presses the designers have found it possible to reduce *a* to zero by positioning the pin A within the thickness of the platen casting. On smaller machines this may make the replacement of this pin difficult but as the total bulk of the assembly is small this restriction is acceptable.

It is normal to find the single toggle mechanism confined to the smaller machines and in consequence the offset loads that are inherent in the design are of a small magnitude and can be readily accepted. However, it is still common to see small machines with only two tie rods. In this case the platen alignment may well be susceptible to variations caused by the offset forces of a single toggle. One well-known German manufacturer has for many years built such a machine with completely satisfactory results, but on close inspection it will be seen that the construction of the machine is extremely robust, and in consequence expensive, especially in the moving platen area.

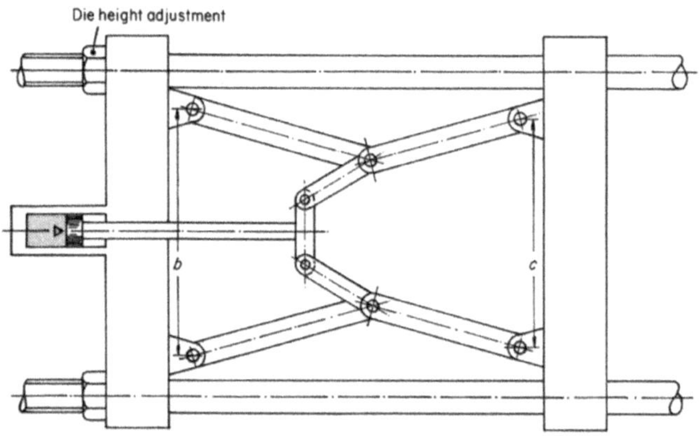

FIG. 2. Double toggle system.

It is noticeable that this manufacturer, in his latest design, has resorted to four tie rods in the interests of alignment and has also changed to direct hydraulic lock, although the single-toggle two tie-rod machines are being offered currently with the later design.

Whilst considering small machines it is important to understand the effects of offset tooling especially in relation to two tie rod designs. It is probably more common to see offset tooling in the smaller sizes due to difficulties in obtaining enough room between tie rods to incorporate complicated runner systems, in consequence these smaller machines are often subjected to loads that are undesirable. These loads can become unacceptable in two tie bar machines and it is imperative, if reliability is to be obtained, that any offset should be arranged so that it is adjacent to a tie bar and not at right angles to the tie bar plane. If this is not possible, and a four tie bar machine is not available, it is worth while investigating the possibility of positioning the offset to counterbalance the twisting forces, if any, imposed on the platen by the toggles. It is true to say that under these conditions the small direct hydraulic lock machine can have definite advantages because of the more even distribution of locking force on the back of the platen and the additional support that the ram may give to this platen.

Unlike the twin toggle method machines it is usual to find single toggle units with equal length main links and therefore harmonic motion of the platen. In general the construction is to have the cylinder beneath the toggle system and this does not allow the designer to take advantage of unequal length links to get extra stroke possibilities. Although there have been one or two attempts to manufacture single toggle systems with some of the advantages of the twin toggle layout, they have not proved advantageous and should only be considered for a very special requirement.

4.2.2 *Double Toggle Systems*

By far the most common toggle system is that using twin links. It is obvious that the side forces discussed on the single toggle are now eliminated from the tie rods being entirely contained within the moving platen. The emphasis on adequate support must now be placed on the method by which the front main link pins are attached to the moving platen and on the stiffness of the platen itself. Today it is rare to find any failures in this area but in the past it was not uncommon to find bolted-on pin anchorage points becoming loose in service, giving rise to early failures due to unequal forces or even, in the worst cases, of over centre movement of one toggle and consequential jamming of the system. This problem will be discussed more fully later in this section.

The forces that have been discussed in relation to the moving platen are of course equally present at the tail plate of the machine. However, as this is not required to move during normal operation of the press, it can be of very much more robust proportions and therefore rarely gives any problem. If, however, pin-mounted pointings are bolted on, the remarks made in the previous paragraph are equally applicable.

As can readily be seen from Fig. 2, showing the twin toggle system, if the two main links are of equal length, movement is severely restricted and in consequence this format is never used in practice. It is now normal to have the front links much larger than the rear and to have the distance (b) somewhat larger than (c). By arranging to have clearance in the pin support system for the links it is possible to get a layout that will allow the main links to swing through an angle approaching 180°. By this means large strokes can be obtained for a relatively short press. It must be understood, however, that no longer is the motion harmonic. Actual platen speed becomes quite complicated for any given ram speed. In the majority of cases this is totally unimportant as the transition from stationary through maximum speed and back to stationary will be totally smooth. It is only when a mid stop position is necessary that the toggle geometry need concern the user. The rule of avoiding the maximum speed point for the halt position should be applied. This point will, however, no longer be in the mid stroke position.

The main links of the system are controlled via a set of intermediate links from the spider fixed to the front of the actuation piston. Although taking very much reduced loads these links and their associated pins have to be just as accurate as the main links.

4.2.3 *Maximum Lock*

In order to get the maximum effect from toggle lock both the main links and the intermediate system must, when fully extended, approach as closely as possible a true straight line configuration through all three associated pins. In the past, controlling this position was often done by stops on the main links, but today it is universal to control the fully extended position of the actuation piston. In order to get maximum locking force this adjustment must never be disturbed unless the necessary measuring instruments are available to reset to the maker's tolerances. If, because of incorrect setting, the toggle should go beyond the 'in-line' condition it may be found impossible to open the press. This is particularly true if maximum lock is being used and the mould is over-pressurised.

The locking force on this type of machine is obtained by stretching the tie bars and unlike either the direct hydraulic or the lock and block, the actual force resisting the injection pressure will increase if the tool is forced open during the injection. In general this is an advantage to the moulder, giving increased control over flash under certain circumstances. However, if the machine is not correctly set it may result in the so-called 'over lock' condition where the pressure available on the opening stroke is not sufficient to pull the toggles from the slightly over centre position that they could have taken up. In this case the press must be re-set to the original maker's specifications. It is not sufficient to reduce the locking force of the machine, as although this may effect an apparent cure, it does not correct the basic fault which will occur again once the press is used at maximum pressure. If, for any reason, 'over lock' condition has been allowed to occur, a practical way to unlock the machine is to place heater bands around the tie rods and allow them to expand. If this is done it should be remembered that the platen must not be moved (just the lock broken) as the guide bushes might well have no running clearance left.

4.2.4 *Toggle Wear and Lubrication*

In order to allow room for the fully opened toggles, the tail plate of the machine will have been deeply recessed and on today's machines this is very often a complicated casting rather than a plain plate. The accurate machining of this part of the machine is of great importance in order to obtain alignment of the complete system. In fact it is this need for extremely accurate machining of the system and maintenance of the design tolerances throughout its life that causes the most concern with a toggle machine.

The reader will understand that due to the forces involved and the absolute necessity to control alignment, the biggest problem of the toggle press is wear. Whereas wear in the main link pins is acceptable to a small extent, any wear in the intermediate links will allow the machine to approach the over centre condition. In order to combat wear all toggle pins and bushes must be adequately lubricated at all times.

It is essential today to have fully automatic lubrication on all but the slowest machines and this system must be arranged so that in the event of failure or blockage it will stop the machine from running. Oil or grease is satisfactory, the latter perhaps resulting in a cleaner machine. The systems are invariably total loss so that excess oil or grease has to be cleaned off at regular intervals. It is common to find catch-trays in the case of oil-lubricated machinery, the oil being returned for filtering and re-use. It is, however, unlikely that any system will be entirely clean and it is advisable to arrange maintenance schedules to contain a cleaning programme. This can have an advantage in that the toggles will automatically be inspected for loose pin caps and lubrication pipes.

Perhaps the greatest controversy between designers in the past has been whether to allow the toggle pins to float fully or to captivate them in one or other of the links. Theoretically a fully floating pin has more bearing area and so is less subject to wear; however, this is only true if it continues to rotate. In practice it will be found that the pin will habitually stop in one position and wear will take place over one area. If this occurs and the pin should move during operations, however slightly, the effective locking force of that toggle will alter. This can be a considerable embarrassment to the moulding shop, resulting in unreliable machinery even to the extent of variations in total locking force between moulding cycles. The fixed pin, although tending to wear at a greater rate, will be much more predictable in practice and has the added advantage of being easier to lubricate.

The end location of the pins is also less important on the fixed type and is normally accomplished by straps that serve to stop the pin rotating. On the fully floating type there will be a tendency for the pin to move longitudinally one way, sometimes with enough force to loosen the end cap. This can be overcome by careful manufacture as it is caused by the effect of the minute 'threads' formed during the final grinding operation. To correct this error the final grind should be carried out so that the 'threads' are handed, i.e. by grinding both ends to the centre.

4.2.5 *Ejector Systems*

The twin toggle machine usually has a great advantage over the direct hydraulic lock when one is considering ejector systems, in that the back of the moving platen is free from all obstructions between the toggle pin housings. Ejector systems with long strokes and multi-positions can readily be accommodated, there being no inter-relationship between platen and ejector movements. There are certain limitations in that there can be no ejectors in line with the toggles and this results in a wide spread of positions across almost the full platen but in one plane only. This plane, being at right angles to the toggles, may cause some difficulties on machines with horizontal toggle systems in that it makes access to the ejector area somewhat more restricted. The choice between the vertical and horizontal toggles may well be governed by this factor, there being no other advantages or disadvantages in either system.

4.2.6 *Mould Height and Locking Force Adjustments*

Mould height and locking force adjustments are usually carried out simultaneously on this type of machine by physically moving the whole of the locking assembly up and down the tie bars. This movement is obtained by rotating the tie bar nuts. The amount of adjustment available depends only on the length of the threaded portion of the tie bars and the amount of slide way for the tail plate on the machine bed. As it is necessary to supply some form of slide way for the tail plate this is often extended to act as the slide way for the moving platen. In consequence of this, only the smallest toggle machines will be found to be without platen slides.

In order to ensure even load distribution between the tie bars it is advantageous to have all the adjustment nuts mechanically interlocked. On small machines this is not always the case. In event of the nuts not being interlocked extreme care must be taken to ensure that squareness is maintained at all times. The manufacturer will have provided some marks by which this can be accomplished. On machines over 50 tonne lock, interlocking of tie bar nuts is advisable and on the largest sizes it will also be necessary to drive the adjustment by some means, because the force needed will be such that manual movement would perforce be unacceptably slow.

In some cases, usually on smaller machines, anti-vibration lock nuts are fitted to the tie bars in front of the tail plate. If these are tightened when the lock is applied it will be found impossible to loosen them with the lock removed. It is therefore always necessary to loosen these nuts *before* a tool change. This is most important, and many examples of the results of disobeying this rule can be seen around the moulding trade.

Whereas, on the direct hydraulic machine and the lock and block, the closing force is a direct function of the hydraulic pressure, the toggle machine depends upon stretching the tie bars. In consequence, the direct reading of the lock applied at any one moment is more complicated. Two methods are in current use, although a third and infinitely more satisfactory system is available on certain machines.

The first method depends on reading the pressure of the oil in the cylinder at the moment when the toggles fully lock. This is usually done by means of slowly increasing the pressure until the toggles just lock. However, if this is done, no account is taken of inertia. This system has the advantage of being simple because the force applied by the operating ram is in direct relationship to the lock applied at all mould heights. The second system is much more accurate, but more difficult to use on the shop-floor since it depends on some mathematical calculations or on the reading of locking curves supplied by the manufacturer. The actual stretch in the tie rods is measured either by a vernier or a dial gauge and this reading is converted into force by means of a simple formula. However, the situation is complicated by the fact that the formula has to be modified for every variation in effective tie bar length. The basic formula does not result in a straight line graph so that the added correction for tie bar length usually takes the calculation beyond the scope of the average tool setter. Press manufacturers, realising this, normally attach the necessary graph to their machines so that accurate locking pressures can be obtained.

Ideally the stretch of each tie bar should be measured and the average taken, but except in the most exacting conditions this is normally not necessary. However, equipment is now available which measures, by means of strain gauges, the forces applied to each tie rod. The results can be displayed so as to give the strain on each tie bar and, in addition, can be used to give the total locking force. In the author's opinion this will rapidly become the accepted method of obtaining locking forces on toggle machines. Because it can differentiate between tie bars, this method has an advantage which can be extremely useful on any type of locking mechanism.

4.2.7 *Tie Bar Refitting*

The removal and refitting of the tie rod in a toggle machine is usually more difficult than on other types of locking systems. Not only must the rod be located correctly in relation to the fixed platen, but the rotational position must also be correct compared to the other rods, to allow the mould height adjustment to operate accurately. Every press manufacturer has his own methods of obtaining these relationships, all satisfactory if carried out correctly, but more care must be taken on the shop-floor if permanent damage is to be avoided than is the case with a direct hydraulic machine.

The problem of what to do in the unlikely occurrence of tie rod breakage is only satisfactorily answered by changing all the rods. This is almost essential in the case of a well-used toggle machine, but unfortunately is rarely acted upon. This results in the early failure of the new rod. The advent of more complicated tooling, especially 'stacked' tools, is today demanding more accuracy from the locking unit and this may well correct this bad practice in spite of its apparent economic advantage.

4.3 Characteristics of a Toggle Lock

Considering the main points of interest on a toggle locking machine, it has:

1. a simple hydraulic system;
2. near ideal inherent motion;
3. small mass of moving parts;
4. complicated highly stressed mechanical components;
5. some unwanted forces on both platens and tie rods; (this depends on machine design);
6. a small pressure area on the moving platen, necessitating thicker plates to eliminate bending;
7. good access for ejector assembly;
8. complete freedom from stroke restrictions caused by mould height;
9. difficult tie rod removal;
10. complete reliance on adequate lubrication;
11. increased effective locking force as the tool flashes;
12. a problem in obtaining readings of the actual locking force.

5 LOCK AND BLOCK SYSTEMS

As the name implies, these systems have two entirely separate elements, one providing the major movement of the platen, and the other supplying the final locking force.

5.1 Means of Obtaining Movement and Locking

The means by which these movements are achieved vary considerably from manufacturer to manufacturer, and in consequence the mechanical and hydraulic systems are too varied to be categorised under a single heading. However, in every case, there will be two distinct systems, the operation of one having to be completed before the initiation of the second. This fact alone means that the system will be inherently slower than the single toggle configuration. In order to overcome the problem, designers have concentrated on speeding up the separate elements to the extent that in certain cases the overall speed of movement is very comparable to that of a toggle machine.

In general, the movements are obtained by systems very similar in operation to those already discussed, but which in each case have the advantage that only one function is required from that element. This allows considerable freedom to the designer to obtain high speeds with good control and still keep unwanted forces at a low level.

Basically the means of obtaining the two motions can be combined in three ways: hydraulics/hydraulics, hydraulics/mechanical and mechanical/mechanical, although the latter is not used on any currently available machine.

5.2 The Double Hydraulic Machines

When considering the double hydraulic machines it will readily be understood from Figs. 3 and 4 that, in every case, it is necessary to interpose some form of mechanical block to isolate the small movement cylinders from the locking pressure applied by the large cylinder, thus adding a third motion to the system.

The movement cylinders are relatively small and in consequence the oil usage is low, allowing compact valve assemblies and comparatively easy control of speeds. Since the whole system can be compared to the direct hydraulic locking unit, but with a mechanical block taking the place of the prefill valve, the limitations of platen speed control still apply, but there are certain advantages to be gained by having considerably reduced reciprocating weights.

Many layouts have been used and the methods of applying the 'block' vary widely from machine to machine, three of the most common being illustrated in Figs. 3 and 4.

In the example shown in Fig. 3 the main movement is obtained by a number of small cylinders (one is shown) and the moving platen has attached to it the blocking bars. The more of these bars that are fitted, the more even will be the load distribution, and in consequence the lighter will be the platen. These bars move through the tail plate assembly of the machine and, when the tool is shut, take up a position such that a sliding block may be interposed between the ends of the bars and the pressure plate.

This block may be constructed as a single plate that rotates through a small arc to cause holes drilled in it to become out of line with the bars, or it

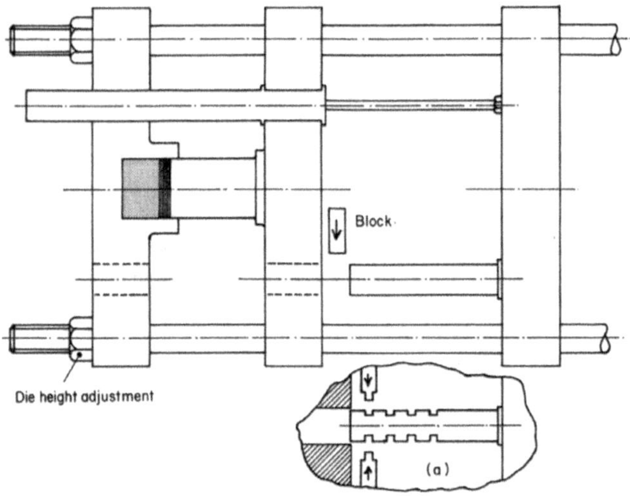

FIG. 3. Double hydraulic lock and block system. Inset (a) shows lock and block machine with variable length blocking bars.

may be a series of sliding blocks as shown. In any case, safe operation must be ensured by sensing before the next stage of the lock is applied.

When the blocks are fully applied the main locking force will be obtained by pressurising the large short stroke cylinder. It will be seen that this cylinder need only be a single acting type and indeed this has been the case in many examples. However, if a double acting cylinder is used, longer life of the unit can be expected as running clearance can be increased for the block movement. It must be pointed out that if this cylinder is of double acting form, mould breathing for thermosets becomes a very easy operation.

It can be readily understood that this locking system imposes no unwanted force on the platen and that the motions obtained are similar to the direct hydraulic lock; however, as the blocking bars are of constant size, it is necessary to arrange some exterior adjustment to accommodate differing tool thicknesses. This is usually achieved by adjusting the effective length of the tie rods in the same manner as on a toggle system. The press stroke is, therefore, independent of tool thickness. Ejector space is only restricted by the blocking bars. In the larger size of press, this system probably gives a most satisfactory ejector configuration. However, care must be taken to ensure either that the platen is thick enough, or that the blocking bars are numerous enough to ensure that no deflection can occur when small or off-set tools are in use.

In order to overcome the added difficulties of fitting adjustable tie rods, a number of machines have been manufactured using variable length blocking bars. As can be seen in the inset diagram in Fig. 3, these usually consist of circular section bars carrying grooves into which the blocks fit. These blocks now invariably consist of a pair of collars which are hydraulically moved into engagement with the bars. It must be understood that, although this system allows for variation in tool height, it is not infinitely adjustable and the movement obtained by the main locking cylinder must now be able to accommodate the full distance between the grooves in the blocking bars.

Accurate positioning of the collars above their respective grooves, before allowing them to enter, is now essential if damage is to be avoided. Various methods of achieving this have been attempted, not all of them infallible. The relative motion is seen in the inset diagram in Fig. 3. It must be understood that the pressure plate is in effect fully floating and has to be correctly positioned with respect to the moving platen before the blocks are applied. This may be relatively easy on a large, slow press, but when inertia forces are high, as is the case in the modern high speed machine, this location becomes progressively more difficult to achieve with any sort of reliability.

It will also be seen that this system, although showing all the other characteristics of the previous one, now limits stroke availability with increasing tool thickness.

Figure 4 shows an entirely different approach to the same problem. In this case the designer has elected to accept the stroke/tool thickness limitation and to accommodate the adjustment by means of a central threaded pressure plate which is allowed to travel inside the main locking cylinder during tool opening. In the tool-closed position the aperture at the front of this cylinder is obstructed by a swinging plate; hence, locking pressure can be applied via this plate. If the limitations on stroke and the restriction imposed on the ejector area can be accepted, this system is by far

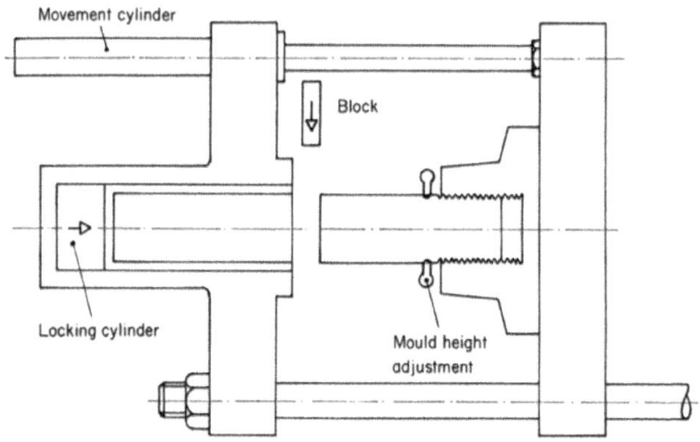

FIG. 4. Lock and block system with threaded central pressure plate.

the simplest and therefore probably the most reliable of the double hydraulic lock and block type.

5.3 Widely Used Systems

Perhaps the most widely used lock and block systems are those combining mechanical stroke movement with final hydraulic clamping.

The already described, near perfect motion of the toggle system can, in this case, be used without the need to resort to large bearing areas in order to combat wear. The toggles are only used for movement and do not in any way contribute to the locking force of the press; they can, therefore, be made light. In fact single toggle systems have been used in quite large presses. All the advantages of speed and smooth movement can be used to the full without the supercritical need for accurate alignment of the straight toggle. The only load on the toggles, other than that imposed by acceleration is that of direct compression during final lock. The geometry can be arranged so that these loads can be accommodated with the toggles in the straight line position. Rubbing forces on the pins are low and in consequence the main disadvantages of the straight toggle are not present in this design, so that lubrication can be simpler and the danger of incorrect operation due to wear absent.

Full advantage of stroke-gaining geometry (see Section 4.2.2) can be taken without weakening the system unduly. As a result, tool movement speeds can usually match those of the larger, direct toggle machines, even with the added time necessary to energise the final pressure cylinder.

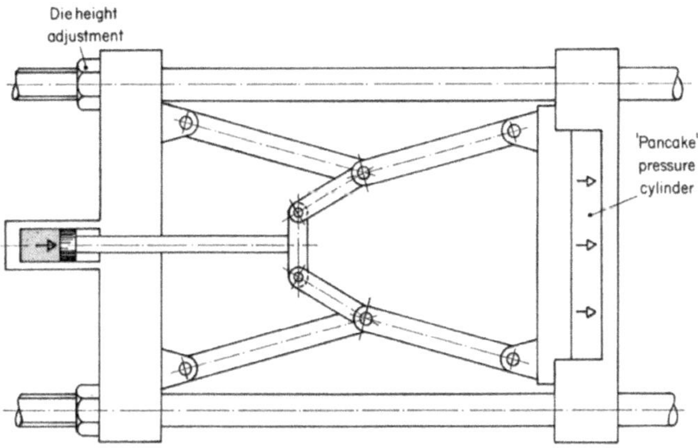

FIG. 5. Lock and block system with the pressure cylinder in the moving platen.

The layout is usually one of two types, as shown in Figs. 5 and 6, the only major difference between manufacturers being the method of obtaining mould height adjustment and the actual location of the pressure cylinder.

In the system shown in Fig. 5 mould height adjustment is obtained by alteration of effective tie rod length, and the locking cylinder is attached to the moving platen. In the system represented by Fig. 6 the adjustment is by single central screw located in the tail plate. The final pressure cylinder is also to be found at this end of the machine.

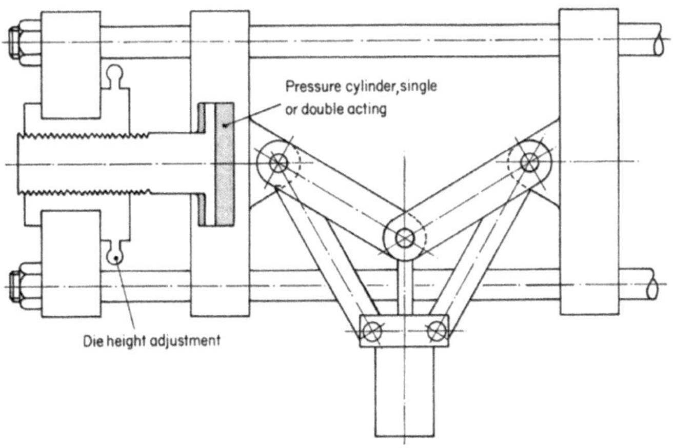

FIG. 6. Lock and block system with the pressure cylinder in the tail plate.

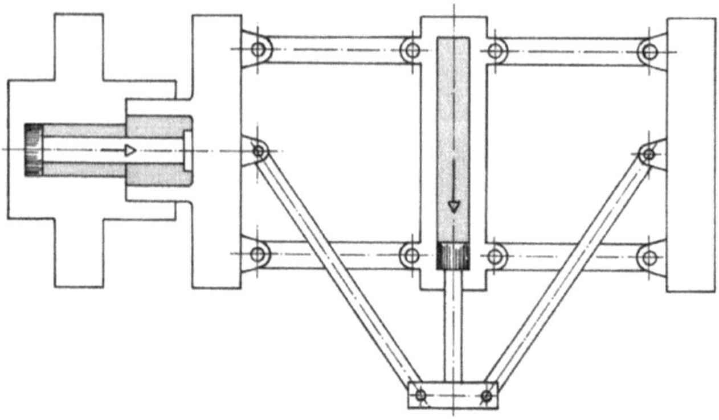

FIG. 7. Compact lock and block system.

Figures 5 and 6 show a single and a double toggle system but it is not intended to imply that these are the only configurations; they are only used as examples to illustrate the basic layout. Figure 7 shows a further layout that has been effectively used to give a more compact press.

The toggle/hydraulic machine is to be found in many sizes throughout the world. One well known UK manufacturer has produced small machines of this type for many years, a development which has resulted in probably the most advanced general purpose locking unit available today.

Figure 8 shows a machine which utilises a single toggle system with an underhung actuating cylinder. Instead of the usual short stroke final clamp cylinder, a much longer double-acting cylinder is used. Mould height adjustment is facilitated by the adjustment of the screwed rod extending from the rear of the final clamp cylinder. This rod serves two purposes, firstly, the positioning of the main piston, and, secondly, the reduction of

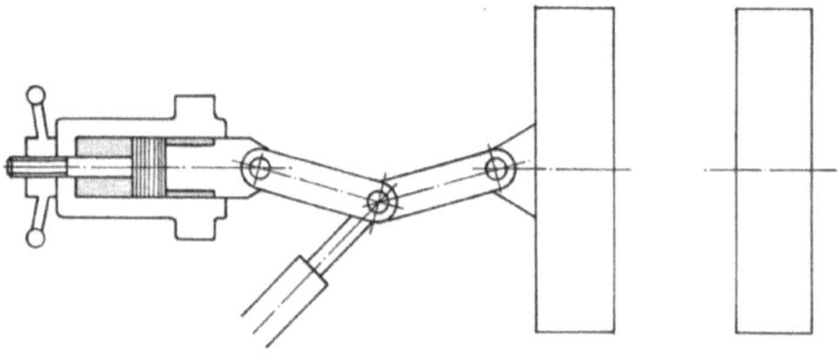

FIG. 8. Useful general purpose locking unit.

the volume of oil contained in the cylinder. This latter fact allows this machine to compete directly with similar sized toggle machines in speed, and the general layout allows for maximum versatility. It will be seen from Fig. 8 that not only is mould breathing for thermosets available, but that also the distance that the tool opens during breathing can be accurately controlled.

One further example of toggle/hydraulic lock and block units is the use of four small 'pancake' (clamp) cylinders fitted to one or other end of the tie

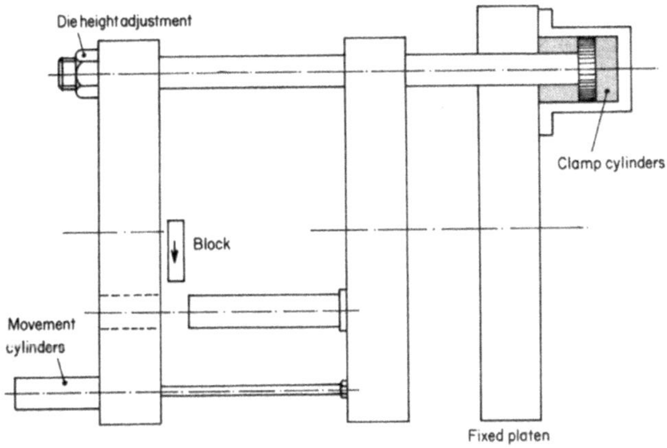

FIG. 9. Lock and block system with clamp cylinders on the tie bars.

rods. Figure 9 shows the general layout. The advantages are obvious: there is no ejector obstruction and it is a relatively short press. Mould height adjustment is invariably carried out by altering the effective length of the tie rods, and in some cases this adjustment is situated at the same end of the rod as the 'pancake' cylinder.

The remarks previously made on toggle speed and position control apply to all machines using toggle systems as the main movement mechanism. The difficulties encountered in obtaining, for instance, a mid-point stop position are not overcome. However, it is true to say that die sensing can be made much more accurate, and as long as the machine setting is carried out correctly, the protection offered by this type of lock to the tool is almost the best available.

Other examples of mechanical movement/hydraulic lock machine have been available from time to time. Probably the best known was the type using an Archimedean screw for the initial movement. This proved to be very reliable and quite fast in operation, but had various lubrication and bearing problems if incorrectly maintained. Basically the toggles shown in Fig. 9 were replaced by a long screw around which a nut was fitted, this nut being driven either by an electric or a hydraulic motor. This motion continued until the tool faces came into contact; then, final locking pressure was applied by means of a direct-acting short stroke hydraulic cylinder.

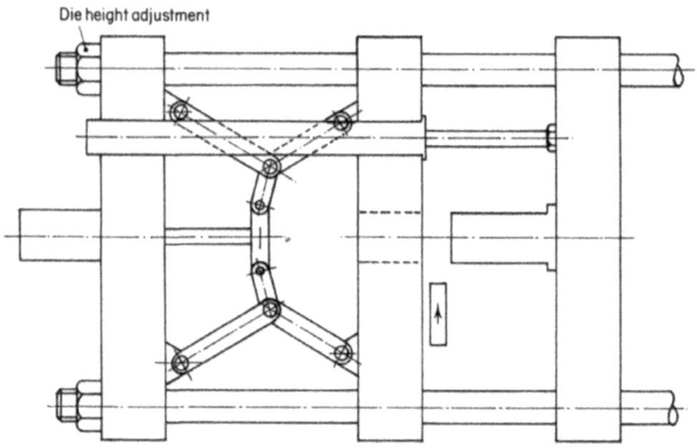

FIG. 10. Reverse lock and block system.

Many examples of this type of machine are still to be found running satisfactorily, but, to the author's knowledge, machines using this principle are not manufactured at the present time.

The final lock and block clamping system, illustrated in Fig. 10, shows a press in which the operations are reversed from the more normal sequence.

The long movement is obtained by hydraulic cylinders, these being blocked off by means of blocking rods and a swinging plate. Final lock is applied by means of strong toggles with a very short stroke. Mould height adjustment is performed by altering the effective length of the tie rods. Unlike the other machines considered in this section, the locking force is not limited by the hydraulic cylinder pressure, but, like the straight toggle machine, will increase as the tool is forced open by injection pressure, the locking force being applied by stretching the tie rods.

5.4 Mechanical Lock and Block Systems

Although very few examples are to be found, the mechanical/mechanical system is not to be ignored. Two forms have been used to good effect, the first being that of an Archimedean screw, to give the main movement, coupled with a short stroke toggle, to give final locking pressure.

The second type was designed in an attempt to overcome the inertia forces of a very large straight toggle machine, and consisted of two sets of toggles set back to back, the front set, carrying the moving platen, being very light in construction and designed to go just to the 'in-line' condition, the final lock being applied by much larger short stroke links.

Neither of these two systems proved either reliable or popular with the moulding trade and are now only of academic interest. They serve to demonstrate two attempts to solve the problems encountered in making a machine outside the size range of that normally made by a manufacturer. In one instance the manufacturer needed to increase his stroke range and so used an Archimedean screw; in the other, a toggle machine manufacturer needed to make a much larger machine than normal.

5.5 Characteristics of the Lock and Block System

Because of the number of permutations available it is not possible to sum up the advantages and disadvantages of the lock and block type of machine. However, if the machine has a final closing system of the hydraulic type, the locking force will be governed by the actuating cylinder. If, on the other hand, the final lock is supplied by the toggle, the remarks on tie rod stretch and locking pressure calculation apply. Likewise the method of main motion will govern the characteristics of this motion and the effects will be identical to those described in the relevant single operation machine.

There will always be two and may be three distinct motions in any lock and block machine and the effectiveness of the final unit will depend to a great extent on how smoothly and quickly the designer is able to obtain the transition from stage to stage.

6 OTHER TYPES

In this section the author has no intention of describing fully the operation of each type, and indeed the following examples only cover a few of the numerous systems that have been tried over the years.

At one time it was considered unnecessary to have more than one hydraulic cylinder on a small machine, this cylinder performing the two main functions of closing the mould and injecting the material. Two examples have been chosen, namely plunger injection and screw.

The first, plunger injection, used a system of toggles and obtained the necessary motion by use of cam plates (Fig. 11). These plates had overstroking slots machined in them to allow the stroke of the hydraulic cylinder to continue, thus operating the injection plunger. Tool thickness adjustment and final locking force were accomplished by altering the effective length of the tie rod, but the stroke was fixed. By means of stroke-

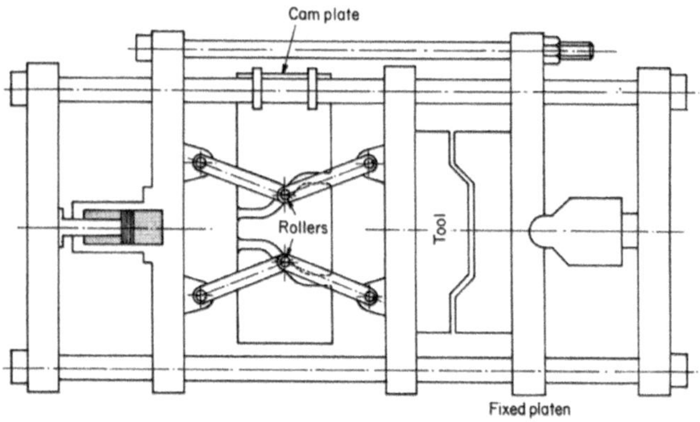

FIG. 11. Single-acting toggle-operated injection moulding machine.

dependent switches injection pressure could be varied independently of locking pressure but the speeds were common to both functions. Die sensing was controlled by pressure, the total locking force not being applied until the rollers had travelled almost completely up the cam tracks. As would be expected, wear (especially at the rollers) caused considerable running difficulties and although satisfactory for some articles, the lack of control did not allow this machine to be called a general purpose one.

The second system was even simpler in that a direct acting hydraulic cylinder was used, moving the platen only (Fig. 12). The cylinder was arranged to have enough stroke both to close the tool and push the closed tool onto the injection barrel and then to continue injecting the material by forcing the barrel over the fixed screw. On opening, the 'fixed platen' was moved away from the barrel by springs and the barrel moved forward by the rotation of the screw. As will be understood, the balance between moulding area and injection force was very critical and although many modifications in the way of balancing cylinders and shot weight control devices were fitted, both by the manufacturer and by the users, the basic limitations to this design are unavoidable. If care is taken with tool design and material selection, satisfactory articles can be manufactured and malfunctioning is unlikely due to the absolute simplicity.

The general layouts that have been considered so far consist of the platens being placed between the locking unit and the injection cylinder. This is not the only accepted configuration and indeed great emphasis has been laid by certain manufacturers on the advantages of having the platens

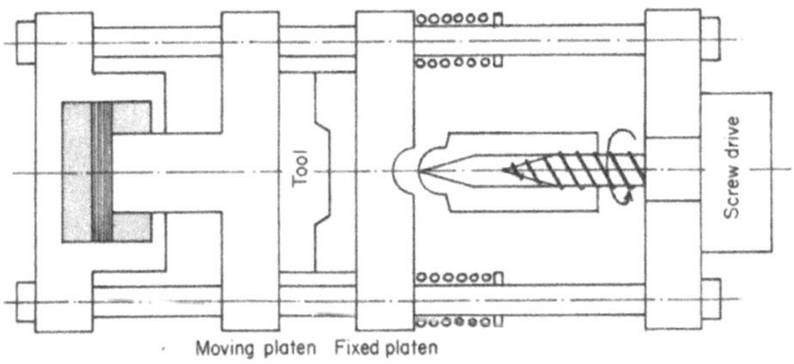

FIG. 12. Single-acting hydraulic injection moulding machine.

at one end of the machine, overhanging the bed, so that easy removal of the moulded article by gravity is possible without obstruction.

In this instance it is universal to place the injection cylinder in the middle of the machine and to have the locking unit at the remote end from the platens that it operates. Certain disadvantages immediately become apparent: the total length of the tie rods is considerable and the overall length of the machine will be great compared with the normal configuration since access to the injection cylinder must be retained. However, the most important point to consider is the effectiveness of the platen support. The moving platen is now at the end of the tie rods and must have adequate support if undue stresses are not to be imposed on these rods with the fitting of heavy tools. When considering such a configuration the method of applying the locking pressure is immaterial; the important area to concentrate on is the stiffness of the platen support and the effect of large tools upon this support. Two additional advantages are gained by this configuration in addition to the previously mentioned access. These are the total freedom for ejector layout, the back of the moving platen being free of all obstructions, and the easy removal of this platen for tool fitting.

In order to complete the picture of locking units it is necessary to mention at least two of the very many special machines that have been manufactured.

The original use of an Archimedean screw was to drive the operating nut to the 'stalled' position, so stressing the tie rods. This motion was developed from vertical blanking presses and for small forces was used for many years with good results, even when driven by a direct current electric motor. Many such presses were in use as recently as the late 1950s but none, to the author's knowledge, was of greater locking force than 12 tonnes.

The second special machine originated in a moulder's maintenance shop and is essentially a method of obtaining high locking forces with very short strokes. Using what is, to all intents and purposes, a normal toggle layout, the operation of the platen is by means of a cam replacing the toggles, the rotation of which is effected through a worm gear from a reversible hydraulic motor. The platen is fitted with a suitable track and its return movement obtained hydraulically, all the running gear being encased in an oil bath. Several of these machines have been observed working by the author and although no maintenance records were at that time available the smoothness and speed with which a stroke of a few inches was accomplished was very impressive. The calculated locking force is 250 tonnes over an area of approximately 24 in^2. Although only of academic interest this example serves to show that it is not necessarily correct to accept machinery that is currently on offer, if long-term production runs are envisaged.

7 CONCLUSIONS

The choice of which unit to buy is still unanswerable, and perhaps always will remain so, but luck plays no part in the purchase of a machine. Careful calculations of what is required, with regard to size, speed, life expectancy and, probably most importantly, control, can reduce the choice considerably. Check that the possible defects in operation of the various types will not embarrass the machine operator on the shop floor. Ensure that the manufacturer has done all in his power to minimise these problems at the design stage. Inspect 'your' machine running before delivery and ensure that it is as smooth in operation as it should be, even under the most adverse conditions. Check the operations of all functional controls and the safety of the guarding systems. Above all, do not accept roughness that, it is said, will disappear as the machine 'runs in'. It will be with the machine forever.

Lastly, the correct function of the locking unit is very dependent on good maintenance, and, whereas faults on the injection end will immediately become obvious by bad mouldings, the effects of wear on the locking end tend not to be noticed until it is too late to correct, except by complete refurbishment.

Chapter 3

MOULD DESIGN AND MANUFACTURE

J. NIGHTINGALE

Consultant, Welwyn Garden City, Herts., UK

SUMMARY

Mould design and manufacture form important links in the chain from the inception of an idea to the mass production of an article. The other factors involved in an injection moulding must be taken into consideration if the best results are to be obtained.

Generally, every mould is different from all others according to circumstances, being composed of features selected to give optimum production. Therefore mould design considerations have been divided according to these features.

The mould-making section reviews the current state of the art, with references to recent techniques and developments including metals for moulds, machining methods and mould finishing.

1 INTRODUCTION

Moulds are essentially 'one-off' items, even if two or three duplicates are made to speed production or to spread risks. This has meant that innovations have resulted mainly from a limited amount of effort and hence have been limited in scope. Up to a few years ago, much of this effort originated with raw material suppliers, as part of their development work to improve the quality and speed of production of injection mouldings.

Many recent developments are the result of extensions and combinations of principles evolved many years ago but now made viable by improved mould making techniques and better process control.

Although a mould is a physical entity, it forms a link in the chain from raw material to finished article. At the same time it is a collection of features that are distinct but related not only to each other, but also to other links in the chain. Most of the sections of this chapter, therefore, are classified according to mould features.

2 INTERDEPENDENCIES

2.1 Shrinkage

'Good tool design starts with good component design.' This is one of the many modern aphorisms that reflect a growing, but long overdue, awareness of the many factors that influence the making of a moulding. A good example of interdependability is shrinkage, which is a perpetual problem for all moulders and mould makers.

There are at least ten variables connected with the mould, the moulding conditions and the material which influence each other and must be taken into consideration every time. In addition, there are others that are significant only in special cases.

Most of the variables that affect the overall shrinkage of all thermoplastics are listed below, and since overall shrinkage is being considered, that is, the dimensions of the final article, post-moulding influences are included.

2.1.1 *Factors Influencing Moulding Shrinkage*

Moulding shrinkage is dependent upon the following factors:

(a) Mould temperature.
(b) Cylinder temperature.
(c) Injection time, pressure and speed.
(d) Type of polymer.
(e) Grade of polymer.
(f) Presence of additives, e.g. fillers, pigments and lubricants.
(g) Length and section of sprue, runner and gate.
(h) Position and type of gate.
(i) Position in the mould, including distance from gate.
(j) Shape of cavity, including thickness of sections.
(k) Direction of flow in the cavity.
(l) Presence and characteristics of any inserts.
(m) Type and condition of injection moulding machine.

Post-moulding influences include:

(n) Stress relieving, by time and/or annealing and/or humidifying.
(o) Humidification—expansion, i.e. negative shrinkage especially applicable to Type 6 and 66 nylons.
(p) Machining operations.

2.2 The Need for Consultation

Best results in the designing of a mould will be achieved only when these variables are borne in mind, together with the production rates required; the minimum acceptable quality in the moulding; the surface character of the finished component; the service requirements from the component; and the post-moulding operations, including machining, assembling, decorating, packaging and transport.

It is generally recognised that the best compromise results from one or more meetings of all the interested parties, namely the end user, component designer, tool designer, tool maker, moulder, and raw material supplier.

Some roles may be vested in the same person and this can be advantageous, particularly in the cases of the end user and component designer, and the tool maker and moulder.

2.3 Mould Proving

Mould makers should have ready access to moulding machines for mould proving purposes but many still rely upon injected wax or on silicone rubbers.

With the injected wax technique, a hard, low-melting wax is poured into the cavity (or forced in from a simple 'injection' cylinder by a hand- or air-operated plunger) via the normal sprue and runner system. Because of the long set-up time for the wax, the speed of filling is not normally critical.

Silicone rubbers can be used either as two-component mixes, which are poured in and allowed to set, or as putty, which is pressed into the mould manually. The latter technique is useful when only part of the mould needs checking; the silicone putty can also be re-used almost indefinitely. The silicone compounds are used because of their ease of release from the cavities, their flexibility and their toughness.

However, no one knows with certainty the properties of a plastics moulding until it has been made, and hence the modification to the mould, moulding material or moulding conditions which will be necessary. Often a simple moulding produces unforeseen difficulties. In many cases, features of uncertainty can be resolved, with saving in time and money, by the manufacture of a prototype cavity (see Section 6.3).

3 MATERIAL FEED FEATURES

3.1 Nozzles

3.1.1 *Nozzles for Single Cavity Moulds*

Although nozzles, apart from hot manifold systems, do not form part of a mould, they are always a factor in tool design. Whenever a single cavity mould is considered, the type of nozzle used is linked to the type of gate that

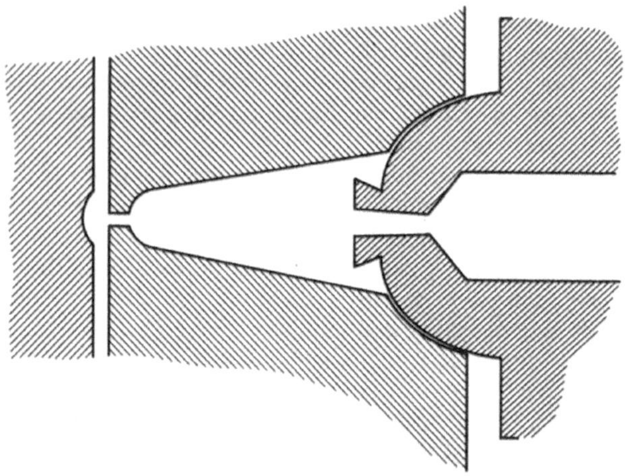

FIG. 1. Hook nozzle used in conjunction with a reverse taper sprue.

is employed. With a single-cavity thin-walled component that needs central gating, a pin-gated reverse taper sprue, coupled to a hooked nozzle, may be the answer in those cases where the tool cost is important (see Fig. 1).

The alternatives to this are a three-plate tool (which would be costlier) or a sprue gate (which may have technical and/or aesthetic objections).

The disadvantages of the system shown in Fig. 1 are: (i) a longer cycle owing to the time taken for the injection unit to make contact with the sprue bush (the nozzle retraction occurs during the mould opening time); and (ii) the need to remove the sprue each cycle. Sprue removal could be carried out either by mechanical means, or by using the machine operator—that is, manually.

3.1.2 *Extended Nozzles*

For sprueless or short sprue moulding, extended nozzles would be used similar to those employed in hot manifold systems. These are discussed in more detail in Section 5.1.

3.2 Sprues

Two aspects of sprue design have been covered in Section 3.1. Other points should now be considered.

3.2.1 *Curved Sprues*

The axis of a sprue need not be straight and this gives an alternative when a single impression moulding needs gating on an edge, but where the

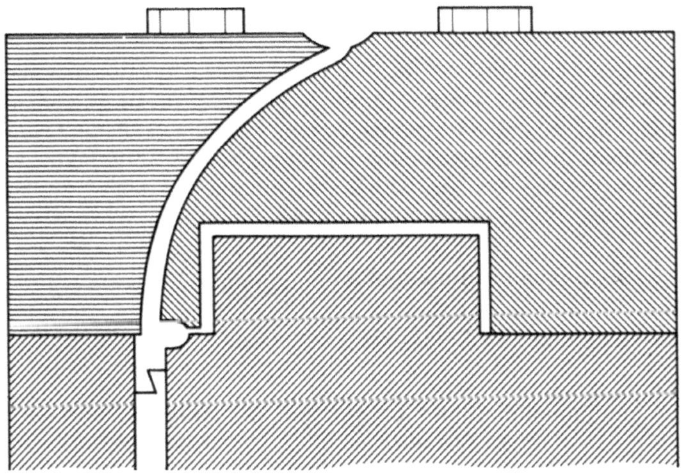

FIG. 2. A curved sprue system.

projected area is centred substantially on the axis of the injection moulding machine. There is a choice of three arrangements: (a) off-set the injection unit; (b) make the mould a three plate tool; (c) use a curved sprue.[1] The last of these is illustrated in Fig. 2. However, points to consider are as follows:

(a) A dearer than conventional tool results when a curved sprue is used.
(b) If a sprue gets stuck in the mould then more time may be needed to clear it (compared to a mould with a straight axis sprue).
(c) The fixed half of the mould may be fairly deep, hence a large mould height may result; this may call for a machine with a higher than necessary lock and shot capacity.

(d) The moulding material must be sufficiently flexible to pull easily from the sprue, but must be rigid enough to have a good anchor to the sprue puller.

(e) The possibility of a longer cycle than with a three plate mould exists, since the factor determining the cooling time would be that for the thick end of the longer sprue.

3.3 Runners

With injection moulding machines giving faster but programmed filling, and the need for more efficient production, a more critical examination is being made of the size and layout of runners.[2,3]

3.3.1 *Calculation of Runner Size*

It has been suggested that the size of a runner where it feeds a cavity via the gate should have a diameter 1 mm larger than the maximum wall thickness.[2] Sprues and main runners feeding sub-runners should maintain cross-sectional areas at each division of flow. These 'rules of thumb' work well for average subjects, and with average injection moulding machines.

A more exact value can be obtained from the following formula:

$$\text{pressure drop (kgf/cm}^2\text{)} = \frac{QLU}{38{\cdot}5r^4} = \frac{QLU}{2{\cdot}4d^4}$$

assuming acrylic at 260 °C with shear of 1000 sec^{-1}.

Where

Q = rate of injection (cm^2/sec): assume 20
L = length of runner (cm): assume 20
U = viscosity of plastic (poise): assume 1400
r = radius of runner (mm): assume 3
d = diameter of runner (mm): assume 6

$$\text{Pressure drop} = \frac{20 \times 20 \times 1400}{38{\cdot}5 \times 3^4}$$

or

$$\frac{20 \times 20 \times 1400}{2{\cdot}4 \times 6^4} = 180\,\text{kgf/cm}^2$$

i.e. about one-fifth to one-eighth of the pressure available from a typical injection machine.

It is interesting to note that if the runner diameter is increased from 6 mm to 8 mm the pressure drop is reduced by a factor of 3·16.

3.3.2 *Late De-gating*

Although the normal procedure with injection moulding is to separate the moulding from the runners as soon as possible, there are many occasions when it is advantageous to leave the mouldings attached. A good example of this is a plastic assembly kit, perhaps of more than 100 components, where numbers on the runners identify the part and indicate the sequence of assembly. Having the components attached to the runners allows easier inspection and packing.

Another case where late de-gating is an advantage is in multi-impression 'family' tools, which again can exceed 100 different and often small components. The spray is held by the sprue for inspection, and is then placed on a cropping table where each component falls down a chute into a container for only that component.

A third example is when mouldings are subject to post-moulding decoration. In one case, part of each runner is given an inverted U shape for positive and stable location of the moulding sprays on holding jigs.

3.4 Gates

3.4.1 *Tunnel Gates*

Although the tunnel gate has been used widely and successfully for a number of years, variations from the conventional 'side-gate' type are still relatively unknown.

The normal arrangement for tunnel gating is to have it in the moving half of the mould, with shear action on ejection. However, with some shapes of components, for example a box, it may be advantageous to place the tunnel in the fixed half and shear the gate on mould opening, by the retention of the moulding on the moving half.

For cylindrical components and bobbins that handle delicate filaments, there may be objections to a gate anywhere on the flange edge or winding area. One solution, if the core dimensions allow, is to have twin tunnel gates in the bore, as shown in Fig. 3. This simplifies tool making for single cavity moulds and gives more symmetrical mould filling than if one or more gates are sited on the outside ends of the flanges.

Tunnel gates with a curved axis, dubbed 'winkle' gates (Fig. 4), were proposed several years ago. They were suggested for buttons and similar articles, where the section at the mould split line had a small radius in both halves of the tool. Difficulties were encountered at the time because the method of forming the tunnel involved spark erosion, with a suitably shaped electrode fed in on a radius.

A variation of this system was used successfully with a multi-impression

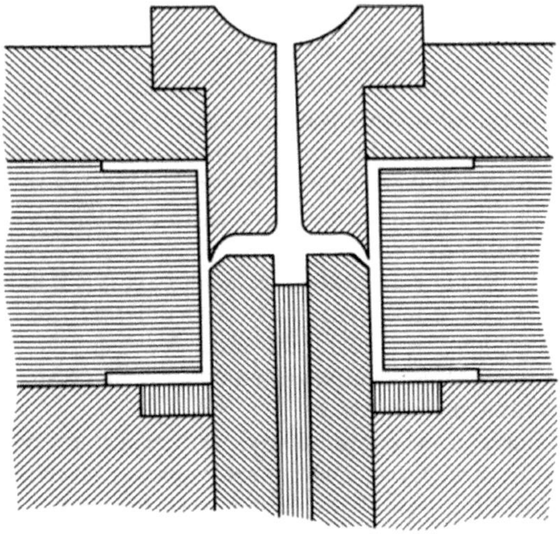

FIG. 3. Twin tunnel-gated mould.

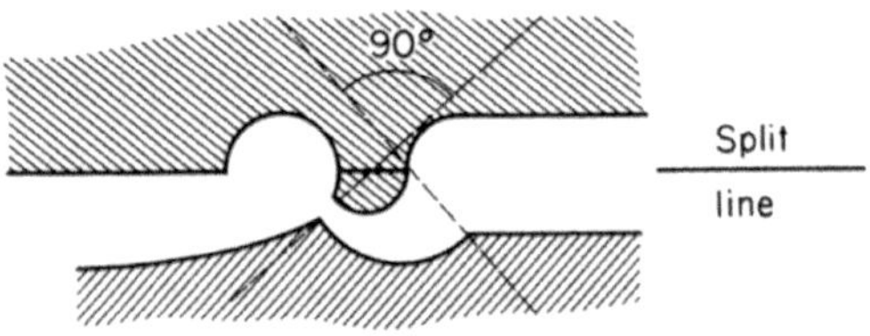

FIG. 4. Curved tunnel or 'winkle' gates. (Scale × 3.)

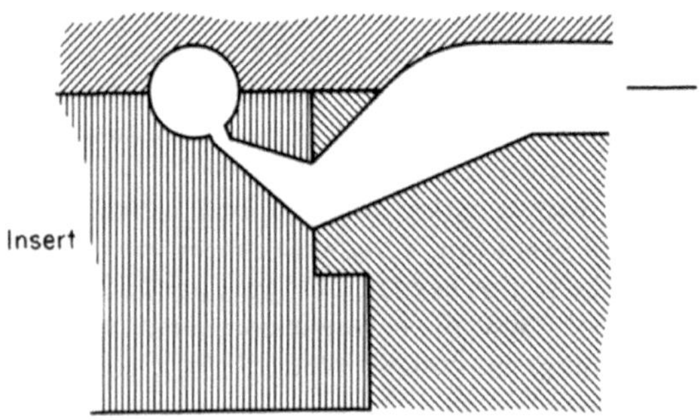

FIG. 5. Three-section winkle gate. (Scale × 3.)

mould with cavity inserts. This allowed the gate to be in three sections, each with a straight axis but at an angle to the next section (Fig. 5). The material used was an unfilled nylon.

A recent alternative method of forming curved tunnel gates is to generate them on the split line of two blocks, which are then inserted into the mould to connect the runner with the cavity.[6] This gives greater flexibility in the design of the gate and might allow for easier fabrication and polishing of part of the cavity if it is incorporated in the gate blocks (Fig. 6).

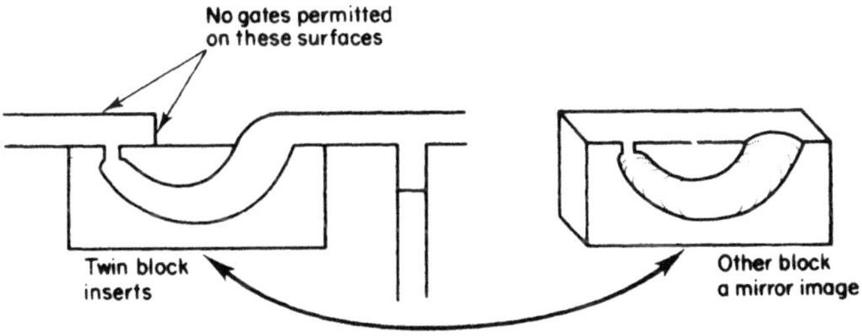

FIG. 6. A method of forming a curved tunnel gate.

With these gates on multi-impression tools it can be advantageous to have wells formed by short ejector pins under the runners, to restrict the longitudinal movement of the runner before the gate has fractured.

Where tunnel gates are severed by tension it must be remembered that the surface finish will not be as smooth as those fractured by shear. Where tunnel gating is not suitable, for example on an external ring-gate, automatic de-gating can be achieved by a shear gate as shown in Fig. 7.

3.4.2 *Flash Gates*

With fibre-filled materials, there is always the risk of warpage due to different degrees and directions of orientation of the filler in different parts of the mould. Where the cavity is substantially rectangular, the pattern of filling can be controlled by the use of a 'flash' or 'film' gate (Fig. 8).

For initial trials the gate would be made of a uniform thickness, but would later be opened up where the rate of filling is slower, to give an even fill. The relatively large area of this type of gate can give the maximum rate of mould filling; this is an advantage for a good surface finish with filled compounds.

3.4.3 *Multiple Pin Gates*

With unfilled polymers that exhibit molecular orientation, multiple pin

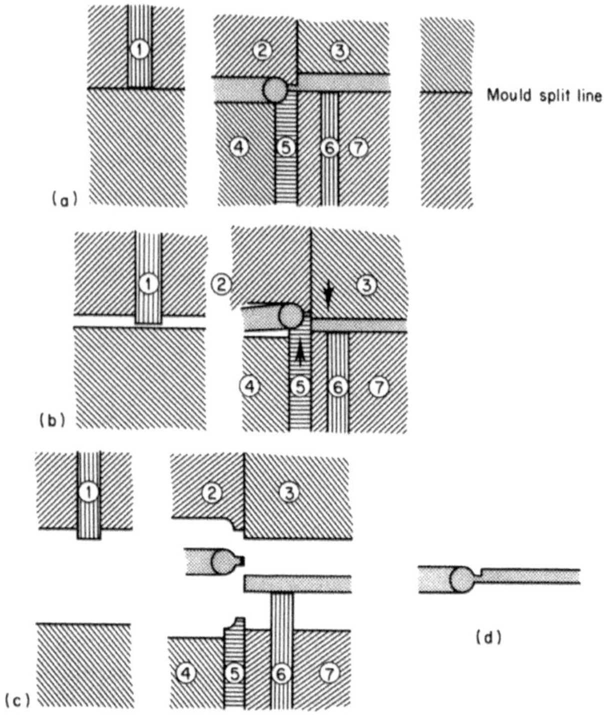

FIG. 7. Automatic de-gating with a shear gate. (a) Mould closed; (b) mould open 2 mm; (c) mould fully open; (d) moulding with 1·5 mm ring gate. 1 = push back pin for punch 3; 2 = fixed half of mould; 3 = slides in 2 by spring washers, 2 mm; 4 and 7 = moving half of mould; 5 = slides in 4 and 7 by spring washers, 2 mm; 6 = ejector pin.

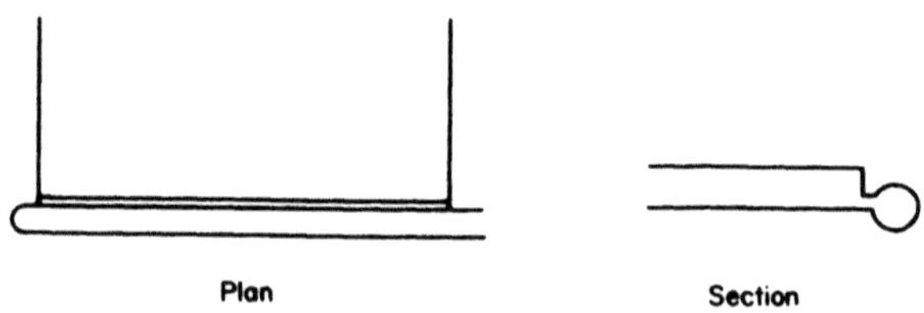

FIG. 8. Flash or film gate.

gates along the length of a component can reduce the risk of warpage. Again, the gate sizes are adjusted to give the optimum filling pattern.

3.4.4 *Gate Sizes*

Gate sizes are still subjects of discussion, even from an attempted scientific approach. The uncertain factor for the formulation of gate size is the viscosity of the injected material. This is very dependent upon shear rates, not only in the gate itself but also in the machine nozzle, the sprue and the runner system.

An adaptation of the formula used for pressure drop in runners in Section 3.3.1 can be used for gate sizes, but as indicated earlier the 'melt' viscosity is an unknown. In all probability, the minimum pressure required in the cavity is that necessary to give an acceptable product.

Since altering the gate is a fairly simple process, the size is usually made to the requirements of the end-user or the de-gating operation, and then opened up according to results and experience if it is considered too small. On multi-cavity moulds, an approximation is obtained with one or two strategically chosen gates, and, when these are satisfactory, the others are opened to similar sizes.

4 RUNNERLESS MOULDS

With the increased interest in fully automatic injection moulding, increased efficiencies and lower costs, there has been an upsurge in recent years in 'runnerless' moulds. Such moulds may have multi-cavities fed directly, or they may be fed via short sub-runners from multiple nozzles.

The manufacture of these moulds has been supported by cheaper precision engineering, new heat insulating and conducting materials, better control of both high and low heats in the mould and of heat and pressure in the injection unit, and better appreciation of mould design.

New literature about 'runnerless' moulds and mouldings appears monthly. Articles which are typical and which have been published fairly recently with reasonable technical details have been under the authorship of Billington, for hot runner moulds,[4] while Csarzar favours the insulated runner systems.[5]

4.1 System Types

Runnerless moulding can be divided into two types: hot runner and insulated runner.

In hot runner moulding the injection machine nozzle feeds into a manifold which is at the moulding temperature of the material. With some polymers the moulding temperature can exceed 300 °C and good insulation both from the machine platen and from the fixed half of the tool is essential.

With the insulated runner system, the manifold block remains fairly cool (e.g. 60–90 °C) and the material flows through a hot core in the middle of the runner. The insulation is formed by walls of cooled injection material between the hot core and the cool manifold; in order to minimise the restriction to flow the runners are normally full round, with diameters two to three times those that would be used if the mould were a 'cold' runner three plate tool.

Although some insulated runner systems have insulated multiple nozzles they gave trouble at start-up due to blocking and nowadays all have externally heated nozzles, or heated probes in the centres of the nozzles.

Recently[5] heated cores were introduced to the insulated runners as a logical extension of heated probes and this has considerably speeded up the start-up procedure.

4.2 Choice of System

Both hot runner and insulated runner systems have their advantages and disadvantages compared with each other and with conventional 'cold' runner arrangements. As with other aspects of mould design, the choice of system is dependent upon many factors including the type and quality of end production and of injection machine, call-off (production) rates, type and grade of polymer, cost, and the experience of the moulder, the mould-maker and the mould designer.

4.2.1 *Advantages of Runnerless Systems*

The advantage of 'runnerless' systems over 'cold' runner systems is the overall cost saving which the former produce. Consider the following:

(i) The shot weight is usually the entire weight of the components. However, in some moulds with large numbers of small cavities, it may be advantageous for each nozzle to feed a sub-runner system with two to twelve impressions. With some small components, the saving in the weight of runners can double the capacity of an injection machine, both in terms of the shot weight and of the plasticising rate.

(ii) The lack of material re-cycled eliminates the risk of contamination and degradation and the cost of re-working.

(iii) Runnerless moulding may give cheaper production (from faster cycles) if one of the limitations in a 'cold' runner system were the time for the runner to freeze.

(iv) A lower locking force per cavity is needed. In some 'cold' runner systems the runner area can equal the projected area of the components.

(v) There is better utilisation of mould area.

(vi) As there are no runners, with their associated ejector pins, there is greater freedom of mould design plus more options for water cooling channels and hence more efficient cooling.

(vii) No provision has to be made for separating components from runners.

(viii) There is the possibility of more consistent quality.

4.2.2 *Disadvantages of Runnerless Systems*

(i) The moulds and instrumentation may cost up to twice as much. This cost can, however, be reduced dramatically if a number of moulds can be operated to use the same hot manifold system. Up to eight moulds under this system are known, so that the cost in this case over conventional moulds is less than $12\frac{1}{2}\%$.

(ii) A longer time for starting may be necessary. However, this could be negligible for long runs, which is the usual case for 'runnerless' systems.

(iii) There is a need for a higher skilled setter, but it must not be forgotten that this could be offset by less unskilled attention during running.

(iv) There is less latitude in gate position and type.

(v) The gate mark may be more obtrusive.

(vi) There are higher maintenance costs for manifold and nozzle heaters for instrumentation, and for nozzle valves where these are used.

(vii) Runnerless moulding is less economic for short runs.

(viii) There is an increased mould height compared with a two plate mould, and moulds using runnerless systems are probably thicker than a three plate tool; this is due to the extra thickness of the manifold block. There have been exceptions to this: for example, where the hot manifold has been placed between the injection cylinder and the fixed platen, with the multiple nozzles going through holes in the platen.

4.2.3 *Advantages of Insulated Runners*

(i) They can be quicker starting than 'hot' runners.

(ii) Construction is simpler and therefore cheaper.
(iii) Less instrumentation is needed (except with heated manifold cores).
(iv) There are lower heat losses from the manifold system, and therefore lower running costs.

4.2.4 *Advantages of 'Hot' Runners*

(i) No separation is needed when starting up (or if cycle is interrupted to remove cold insulated runner).
(ii) There is less risk of degradation on the walls of the manifold, since no solid material is just below solidification temperature (which could have a significant degradation rate).
(iii) Simultaneous filling of cavities is easier to control as the numbers of cavities increases, compared with insulated runners.
(iv) Melt temperature does not have to be higher than optimum in the injection cylinder to compensate for heat losses in an insulated runner.
(v) Hot runners are more suitable for thick section moulding.

Typical applications for runnerless mouldings are millions per year of closures, drinking cups, small jars, small electrical and electronic components, aerosol components, fasteners and cable ties.

5 MOULD COMPONENTS

5.1 Cores

Two consistent problems arise with mould cores, namely undercuts and cooling. Undercuts on the core are normally associated with screw threads. With large diameters such as jar caps, use can be made of gaps in the threads to allow that part of the core to retract on the cap axis, thereby allowing the other parts to move inwards, releasing the cap (see Fig. 9). Temperature control of these parts can be difficult.

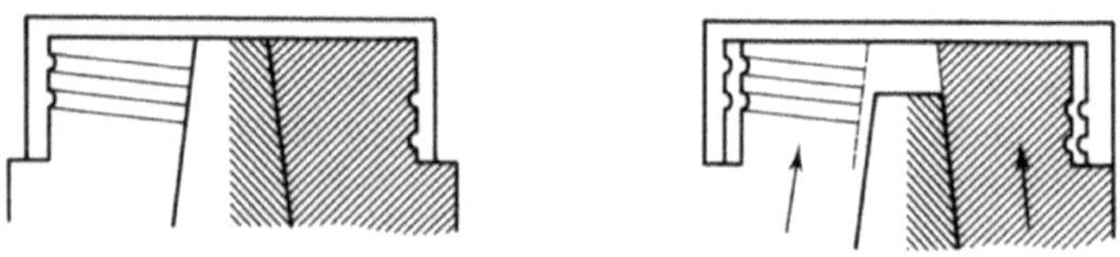

FIG. 9. Method of releasing a threaded cap. Left: moulding clear of cavity; right: moulding being ejected (no thread gives clearance on centre section).

The cooling of thin cores is always difficult and with water or aqueous liquid cooling the precautions which are set out in Section 5.3 should be observed. Recently a pin with high conductivity has been marketed[7] that is placed in a central bore in the core pin, with thermal continuity maintained either with a thermoset resin or a high conductivity silicone grease. The thermal conductivity of such pins is claimed to be 1000 greater than that of copper and such pins replace mould fountains, cascades, etc. Core cooling is achieved by allowing the pin to project into a simple waterway.

5.2 Ejectors

The advent of the 'closed-loop' control of injection moulding has added another factor to mould design, since a signal from the mould rather than time can be used to make a change in injection conditions.

A common method of control uses a pressure transducer in the ejector plate under the head of an ejector pin. The siting of one or more of these ejector pins (which could be used solely for sensing), is all-important.

The filling of the cavity can also be monitored by a heat sensor which detects when the moulding material reaches a certain point or position and can be more sensitive than a pressure sensor.

In some cases a change in injection conditions may be required as soon as a runner system is filled. In this case the sensing ejector would be placed near one end of the runner so that the rest of the runner fills in the time needed to operate the hydraulic valves; such valves change the injection conditions of speed and/or pressure.

5.3 Temperature Control Channels

With the need for more consistent quality in some types of moulding on the one hand and faster production with economy on the other, more attention is being paid to controlling mould temperature.

One mould designer has said that when beginning a mould design, he first sketched the water channels, then the positions of the ejector pins, and finally fitted the plan of the mould cavity over these locations.

The amount of heat carried by the injected material to a mould can be calculated fairly easily. Extracting the optimum quantity of heat in the optimum time is more complicated but formulae have been published[8,9] and used. Theoretical and practical examples have been made recently by the Northern Ireland Polytechnic.[10]

The diameter and spacing of the cooling channels is dependent upon a number of factors, including the shape of the component, the material being injected, the desired quality and the mould metal or metals. A couple of guidelines have been given,[9] stating that the depth of the centre line of the channels should be one to two times the channel diameter, while the pitch between channels should be three to five times their diameter.

When making calculations regarding water cooling, exact data are not normally required. Often, as in the example which follows, simple calculations are adequate to check the order of design features.

If the heat balance, checked before mould design and calculated before mould manufacture, cannot be achieved under the assumed conditions, then alterations to the mould design and/or moulding condition can be made at an early stage, with little loss of time, money and the goodwill of the end-user.

5.3.1 *Water Cooling Calculation*

Assumptions:

Material: GP polystyrene
Material temperature: 200 °C
Shot weight: 100 g
Cycle time: 20 s, i.e. 180 shots/h
Average temperature of moulding on ejection: 100 °C
Mould cooling water in: 19 °C
Mould cooling water out: 25 °C.

Enthalpy of GP polystyrene from 200 °C to 100 °C = 45 cal/g.

Therefore heat input to mould = enthalpy × weight × cycles = 45 × 100 × 180 = 810 kcal/h. Volume of water at 6 °C difference to extract the heat = 810/6 litres/h = 135 litres/h = 0·135 m^3/h. Assuming 10 mm bore to cooling channels (area 0·78 cm^2):

$$\text{water velocity} = \frac{135 \times 1000}{0{\cdot}78 \times 100} = 1730\,\text{m/h} = 0{\cdot}48\,\text{m/sec}$$

This will give turbulent flow in the mould channels and hence maximum efficiency in cooling.

Turning to mould design, we have to make some wide approximations, since exact calculations require a computer program. With the cooling channels at 10 mm below the mould surfaces, and at 30 mm centre spacings, it is assumed that this is equivalent to having continuous cooling over the whole area at a depth of 15 mm.

The average temperature of the cooling water is 22 °C and the average temperature of the moulding is 100 °C—a difference of 78 °C. Assuming that the projected area of the shot is 35 cm^2, the cooling area is 70 cm^2 (two halves of mould).

The thermal conductivity of hardened steel is 0·22 kcal/h/cm thick/cm^2 area/°C. Therefore,

$$\text{rate of heat conduction} = 0{\cdot}22 \times \frac{10}{15} \times 70 \times 78 = 800\,\text{kcal/h}$$

which is of the same order as the 810 kcal input of the injected material.

If the heat extraction had not been sufficient, one can see at a glance where modifications can be effective or ineffective. For example:

(a) increasing the water cooling rate by 3 would only increase the temperature gradient by 2 °C and increase heat transfer by a factor of 80/78;
(b) using brine at 0 °C, the factor is only 100/78;
(c) using case-hardened mild steel would nearly double the rate of heat extraction, and if beryllium copper were used a still greater factor would apply;
(d) placing smaller channels only 5 mm from the mould surface would double heat extraction.

6 MOULD DESIGNS INFLUENCING MOULDING TECHNIQUES

6.1 Vertical Flash Moulding

Although 'vertical flash' moulds were developed more than a decade ago, they are still relatively unknown. An alternative name for the associated moulding process is 'injection stamping'.[23] The process is mainly applied to mouldings of large areas, since the moulding is formed with much lower 'locking' pressures.

6.1.1 *The Principle of Vertical Flash Moulding*

The process depends upon the injection of the required volume of material at a low locking force, immediately sealing off the feed and then advancing the moving half, or a large part of the mould, to complete the filling of the cavity and take up thermal contraction (see Fig. 10). By this method acrylic mouldings weighing 17 lb and having a projected area of 2100 in^2 have been produced with a mould lock of 1300 tons.

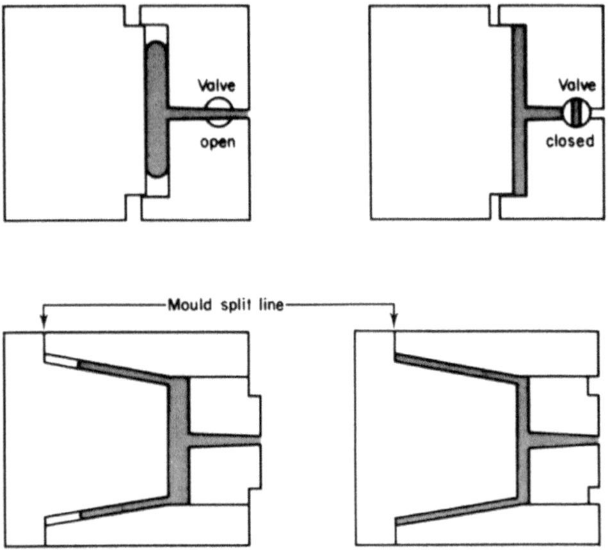

FIG. 10. Vertical flash moulding. Left: volume injected; right: pressure applied.

In order to vary the volume of the cavity there must be a parting line that is in line (or nearly so) with the mould closing direction. This can be either near the normal split line or at the bottom of a cavity on either half, as shown in Fig. 10.

The injection of two materials in sequence into a mould during each moulding cycle[21] allows a continuous high quality skin to enclose a core with different properties. The two materials, which need not be based on the same polymer, must be compatible to avoid mechanical weakness.

6.1.2 *Advantages and Disadvantages of Injection Stamping*

The advantages are:

(i) the ability to make mouldings with one half to one sixth of the normal locking force;
(ii) lower moulding stresses, since the mould filling pressure is relatively very low;
(iii) the ability to start plasticising the next shot as soon as the feed valve has closed.

The disadvantages are:

(i) restrictions in component shape;
(ii) extra mould and machine costs;
(iii) restrictions in moulding materials, for example, it is not applicable to many filled compounds. The process has been used successfully with unfilled acrylics, polystyrene, polythene, ABS and polycarbonate;
(iv) the process is restricted to hydraulic mould clamping, but toggle closing may still be possible;
(v) the possibility of greater mould wear at the vertical shut-off surfaces. With the higher viscosity polymer, this may be very small. Much depends upon component and mould design, and choice of mould metals and treatments.

6.2 'Composite' Moulding

The title is intended to cover the production of injection mouldings, where a part of the moulding is not produced during that cycle.

6.2.1 *Insert and 'Outsert' Moulding*

Insert moulding has been widely practised for many years and the volume of the insert is generally less than the volume of plastic which partially encapsulates it.

Recently a number of examples have gone into production where plastic material has been injected on to a relatively large metal component. In one quarter this has been given the name of 'outsert' moulding.[11]

If a distinction needs to be made between insert and 'outsert' moulding, it is suggested that the former is where the volume of the second material is equal to or less than the injected part.

'Outsert' mould considerations. Features to be considered when designing moulds for 'outsert' moulding are as follows:

(i) Where the metal component extends outside the cavity the mould locking system should be hydraulic (to allow for slight variations in the thickness of the metal). Toggle locking can be used where a silicone sealing ring can be incorporated.
(ii) The positioning of metal sheets is normally by punched holes which are located on pins in the mould.
(iii) Shrinkage considerations are different from those usual in normal moulding. If straight runners are placed only on one side of a plate, there is a strong risk of the plate's bending. Either identical runners

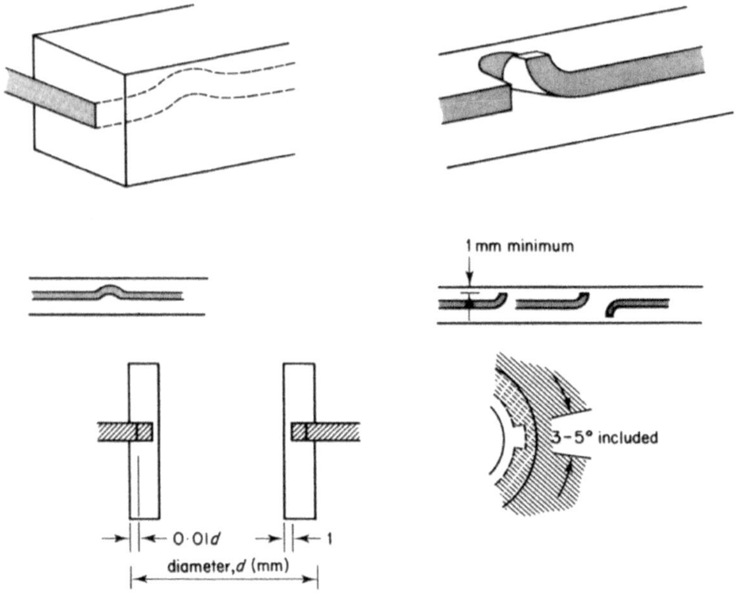

FIG. 11. Methods of obtaining keying using slots or ribs.

should be placed on both sides, or the runner should be sufficiently curved to absorb the shrinkage. It must be remembered that where the plastics material is moulded through a hole in a plate, it will shrink on to the thickness of the plate but away from the sides of the hole. Some form of keying may be necessary to stop any movement of the plastic in the hole. This may be achieved by means of extra small holes outside the main hole that are punched at the same time as the main hole. This may necessitate a flange which in some circumstances may be undesirable. Other possibilities are ribs, or tapered slots that break into the main hole. These alternatives are shown in Fig. 11.

(iv) Where a core passes through a metal plate, scoring of the core with the plates can be avoided by incorporating a moulded edge 1 mm wide (for example) around the hole. This ensures that any metal-to-metal contact will be avoided.

(v) If plastic is going to be injected on the edge of the metal, then the metal must be enclosed completely within a cavity.

(vi) To avoid stress concentrations where plastic comes into contact with punched metal, the edges of the hole should be chamfered.

(vii) 'Outsert' moulding is economic normally only when a number of plastic parts are moulded on to one piece of metal (even if the metal is cropped subsequently to leave one moulding per metal piece). Thus, tools are designed usually as three-plate or 'hot-runner' moulds.

6.2.2 *Sink Mark Elimination*

In some critical mouldings it may not be possible to disguise a sink mark caused by a locally thick section, for example a boss. This marking can be

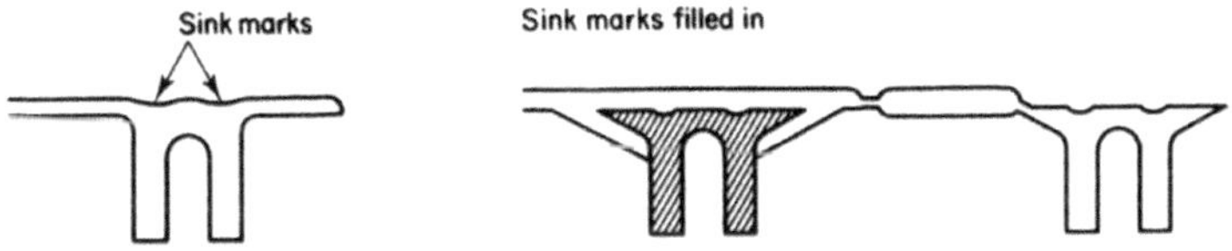

FIG. 12. Inserts moulded from parent material.

eliminated by the use of an insert of the same material made in the previous moulding. As there will be little or no fusion between the new moulding and the insert, mechanical keying is necessary, and since the insert need be no thicker than the main moulding and is moulded at the same time, costs are kept low. The principle is illustrated in Fig. 12.

6.3 Prototype Moulds

Modern knowledge and computing facilities have taken a lot of doubt out of component and mould design but there are still areas of uncertainty due to a lack of detailed data, on, for example, the harmonic vibration frequencies and amplitudes occurring in a plastics power-tool housing at different temperatures.

Often the quickest and cheapest method of obtaining data for obscure problems is by prototyping. The simplest method of obtaining the component in the plastics material is often by machining a component or part from solid, or, where this is not possible, by the use of a prototype mould. A number of examples and methods have been covered in recent years in symposia organised by the Plastics and Rubber Institute.[19]

At the bottom of Fig. 13 are shown two metal components which the

FIG. 13. Components required in prototype quantities. (Reproduced by courtesy of ICI Ltd.)

FIG. 14. Prototype mould using a plaster insert. (Reproduced by courtesy of ICI Ltd.)

manufacturer wished to have made in a plastics material. For evaluation purposes a cavity was produced in plaster (Fig. 14) and this allowed the sample moulding (shown at the top of Fig. 13) to be produced. Prototype moulds are conveniently fitted into a prototyping bolster (Fig. 15).

One of the most unusual methods of making large prototype tools is the use of sprayed metal backed with concrete.[20] Many of the methods of making prototype moulds are also applicable for short run production. It is in this field that aluminium inserts or moulds have their greatest potential (see Section 7.2.2).

FIG. 15. Bolster set used for experimental moulding. (Reproduced by courtesy of ICI Ltd.)

7 MOULD MANUFACTURE

As with design, mould making cannot be viewed in isolation. Parts of the mould other than the cavity surfaces may need careful attention and processes such as hardening and annealing must also be considered. Other instances are the manufacture of hobs for cavities and the preparation of patterns for casting or electro-deposition. The choice of the grade of steel and its treatment will vary according to cost, injected material, length of run and intricacy of cavity.

Although moulds are usually 'one-off', some 'mass' production methods are helping to keep down costs, as in the cases of standard bolsters and insert blanks, and cast metal cavities (pressure-hobbed or electro-deposited) in multi-impression tools.

7.1 Standard Parts

Bolster sets are now used extensively for medium and small moulds. They can be produced cheaper and quicker than 'specials' and, being built on a modular principle, they are very versatile.

Coupled with bolster sets, there are those that have insert cavities precision-machined to standard sizes, with mating insert blanks. Thus, in multi-impression tools, cavities if damaged can be replaced or modified with a minimum of time, cost and interruption of production. This is particularly the case where large numbers of small mouldings of a varied shape and an unpredictable demand (such as ring travellers for the textile industry) are required.

The large-scale production of many ancillaries, such as water-line connectors, makes individual manufacture uneconomic. In addition, instantaneous connectors can save time and cost when changing moulds. When using ancillary hydraulic circuits, the shut-off feature on these instantaneous connectors avoids a considerable mess on the shopfloor.

7.2 Metals for Moulds

Steel is still considered the 'standard' material for injection moulds and in recent years special grades have been developed especially for injection moulds.

7.2.1 *Steel*

Pre-hardened steels are used fairly extensively nowadays in cases in which maximum durability is less important than the risk of distortion on hardening of an intricate cavity. These steels have a hardness throughout of about 32 Rockwell C (300 Brinell), so that they can be machined and given a high polish without the necessity of further finishing. However, in some circumstances it is possible to improve local hardness by nitriding, flame-hardening or plasma arc treatment.

New steels have been developed that are more resistant to corrosion than those used previously. These can be advantageous for the moulding of PVC and other polymers giving off corrosive fumes on decomposition, such as certain fire-retardant grades. Here the choice has to be made between 'special' steels that give good corrosion resistance (with no after treatment) or 'standard' steels that have some form of surface added; this added surface may give superior results.

Another class of steels developed in recent years is those known as 'maraging' alloys. They are mostly stainless steels suitable for mould making and have the advantage that after machining they can be age-hardened at 400–500 °C. This is another of the alternatives for reducing the risk of distortion on hardening.

7.2.2 *Other Materials*

For areas where there is the likelihood of high wear rates (e.g. gates for moulding polymers with abrasive fillers) tungsten carbide or other ultra-hard materials are used. Because of their cost and some limitations in physical properties a minimum of material is used. Electro-discharge is the usual machining method for such materials.

A recent survey showed that moulds made from wrought aluminium plate could be produced for 32 % of the cost of a steel mould with the same degree of polish in the cavity. The life of such a mould is obviously dependent on its conditions of use, but typical runs are in the range of 10 000 to 100 000 off. Thus, if mould costs are a significant percentage of component costs, aluminium must be considered.

7.2.3 *Cast Moulds*

Cast mould cavities are used mainly for mouldings that need an intricately patterned surface and can be taken from a simple model. Cast zinc alloy moulds have been used in the UK for many years but the growth alloy is now beryllium copper processed by a precision casting technique.[12] As with the zinc alloys, cooling channels can be cast in, to give leak-proof labyrinths with smooth bends.

7.3 Mould Machining

7.3.1 *Electro-discharge Machining*

Electro-discharge machining (EDM, spark erosion or burning) is the removal of metal by the melting action of a spark. The removal of the waste metal and heat is by a circulating fluid, usually an oil, but other liquids have been used. The sparking is between the work-piece and an electrode; this is fed into the work at a fixed rate until the required depth is achieved.

This process of machining a wide range of metals has come very much to the fore in recent years, and there are very few mould makers who do not have at least one machine. The process itself is now better understood, and it has been established that the metal removal is primarily thermal.[13,14]

Most EDM work is down a single axis. This is logical since its purpose is to reproduce a cavity, in part or complete, of a punch that represents (in broad terms) the external surface of the moulded article. Its utility is extended considerably when the movement of the electrode moves into other motions. Thus a radiused feed causes the generation of curved holes, the only alternative to which is the use of split blocks which have already been mentioned.

Another example of two-dimensional movement is when the electrode, usually wire, has its movement numerically controlled.[15]

The use of very small orbital movements of the electrode in a plane at right-angles to the feed axis allows one electrode to cover both roughing and finishing.[15] The success of this depends upon conditions that give low electrode wear.

Copper alloys still hold 'pride of place' for electrode materials, but in some circumstances there are advantages in using cast electrodes based on epoxide resins. These are filled either with graphite or with powdered metal and such fillers may be used alone or with the addition of one of the newer carbon blacks that give good conductivity as a result of their particle shape, even at relatively low concentrations.

Work at the Cranfield Institute in the UK shows that electrode wear can be reduced considerably, for example by as much as one quarter to one third, by the use of the optimum electrical pulse. This and other controls to suppress arcing, will give faster metal removal for a given amount of electrode wear, or a better finish and less electrode wear with present-day 'standard' feed rates. The linear value of electrode wear has been found to be greater on corners than on faces.

Where complex parting lines have to be mated it may be found that EDM can be considerably faster than 'spotting and scraping'. Two important factors are the need for a rigid head to hold the lighter part of the mould (usually the fixed half) and the maintenance of a good flow of flushing fluid. Ejector pin-holes can be useful channels to aid both the flow and the distribution of the fluid.

The slightly textured surface of the parting faces can be advantageous, since it will give more even and efficient venting during moulding than many ground faces.

7.3.2 *Numerically Controlled Milling*

Numerically controlled (NC) milling was hailed some years ago as the greatest advance since die-sinking. Experience highlighted a number of unforeseen difficulties, and it is only in the last few years that the process has settled down to its true place.

In numerically controlled milling the motion of the cutter to the work is controlled by motors that receive motivating impulses from programmed punched or magnetic tape. In addition, the feed and cutter speed can be controlled from the tape and in the latest type of machines the actual position of the cutter can be fed back into the control unit to give anticipatory and corrective movements.

Recent innovations include improved accuracy in positioning the cutter and anticipatory changes of feed at the ends of cuts or rapid changes of movement in one or more of the axes.

Where three-dimensional shapes have continually changing sections, as in fan blades, a pattern is normally made to guide the milling cutter. If the change from one section to another is simple, and can be expressed as a mathematical formula, then use can be made of computer graphics coupled to a suitable NC machine.[16]

Many shapes can be defined geometrically so that a change from one section to another can be expressed mathematically in terms far simpler than a series of section drawings. A curved axis in a component usually means the addition of only one more expression, but it would constitute a number of blending radii to a pattern maker. It has been found that components made to computer graphic designs usually work more efficiently in service.

7.4 Mould Finishing

One of the major risks with steel tools that require hardening is the possibility of distortion and, occasionally, cracking. Getting an even state of hardness down holes or, at the other extreme, masking off certain sections, can be difficult.

7.4.1 *Plasma Nitriding*

Most of the difficulties mentioned above can be overcome by the use of plasma nitriding.[17] In this process the machined parts are in a vessel with an atmosphere of hydrogen and nitrogen at a low pressure (0·1–10 torr). An electrical potential of 100–1500 V is applied with the mould parts as cathodes, and a luminous gas discharge is formed. This effect is uniform on all parts of the surface, even into blind holes. The process takes place at 400–600 °C in a semi-vacuum, so that power consumption is low and there is a minimum risk of distortion. Although this process is widely used in Germany and Japan, as far as is known no production facilities exist in the UK.

7.4.2 *Mould Plating*

The plating of moulds has been undertaken almost from the start of moulding so as to give both better corrosion resistance and a higher finish, and the addition of metal can be used to build up worn or damaged parts.

For protection against the effects of PVC moulding, electroless nickel has been favoured for a number of years. It has the advantages of giving a porous free coating and an even layer of metal at the corners.

The latest practice in some moulding shops is to give a 2·5 μm layer of a hard gold alloy. This can be deposited directly, and *in situ*, on to steel or nickel and is then an important factor in maintenance considerations. Unlike other established methods, this technique[18] uses a cotton-wrapped anode, which is dipped into the appropriate electrolyte and brushed over the area to be plated. The mould is connected to the negative power supply and the total amount of metal deposited is gauged by an ampere-hour meter. Areas covered have been up to 1 m^2 while thicknesses have been from a few hundredths of a micron to 0·75 mm. Because of the simplicity and portability of the process, replacing gold films or building up worn parts can be undertaken with the mould still in a press.

7.4.3 *Mould Polishing*

With the possible exception of moulds for optical components, the polishing of moulds is generally done by hand. This is expensive and can constitute up to 30 % of the cost of a tool. Mechanical support in the form of rotational aids has been used for many years; these aids are not, however, effective in certain shaped recesses such as V sections, and currently on the market are tools that give a reciprocating motion which supplements rotary polishing motions.

Where a reciprocating motion is used for finish polishing after a tool has been proved, part of one or more of the proving mouldings could be used as a lap.

REFERENCES

1. Bayer Plastics. E.421-71 B/63216.
2. SMITH, J. B. (1978). Paper 4, *PRI Symposium, 'Mould Design and Productivity'*.
3. PYE, R. (1977). *Injection Mould Design*. PRI Monograph, Iliffe.
4. BILLINGTON, E. (1976). Paper 1, *PRI Symposium, 'Mould Design'*.
5. CSASZAR, E. J. (1978). Paper 3, *PRI Symposium, 'Mould Design and Productivity'*.
6. Bayer Practical Information, 1st January 1976.
7. 'Kor-Kool' Pins: Polymer Engineering and Design Aids Ltd, Hailsham, E. Sussex, UK.
8. WATERS, C. E. (1969). *Modern Plastics*, April.
9. PRASAD, A. (1974). *Plastics and Polymers*, December, 265.
10. CRAIG, J. E. and MILLER, P. P. (1978). Paper 5, *PRI Symposium, 'Mould Design and Productivity'*.
11. Amcel Ltd, Watford, Herts., *Outsert Moulding*.
12. GUBBAY, G. Y. (1977). Paper 2, *PRI Symposium, 'Mould-making 77'*.
13. CROOKALL, J. R. *et al.* (1971). *Ann. CIRP*, **20**(2), 113–19.
14. CROOKALL, J. R. *et al.* (1974). *International Symposium on Electrical Machining*. Bratislava, 18th–19th September, 68–85.
15. CROOKALL, J. R. (1977). Paper 1, *PRI Symposium, 'Mould-making 77'*.
16. WELBOURN, D. B. (1977). Paper 4, *PRI Symposium, 'Mould-making 77'*.
17. BELL, T. (1976). Paper 6, *PRI Symposium, 'Mould Design'*.
18. OGDEN, S. and RUBENSTEIN, M. (1977). Paper 3, *PRI Symposium, 'Mould-making 77'*.
19. *PRI Symposium, 'Low Cost Moulds'*, 24th May 1972 (most papers).
20. GARNER, P. J. (1971). *S.P.E. Journal*, **27**, 17–22.
21. ICI Ltd (1972). *The ICI Sandwich Moulding Process* 2nd ed., ICI Plastics Division, Welwyn Garden City, Herts., UK.
22. *Rubber and Plastics Age*, April 1965, 427 and 429.
23. *S.P.E. Journal*, **24**, 1968, 61–3.

Freedom from patent rights should not be assumed.

Chapter 4

SCREW AND BARREL ASSEMBLY DEVELOPMENT

J. B. SMITH

Account Manager, Plastics Products and Resins Department,
Du Pont (UK) Ltd, London, UK

SUMMARY

This chapter briefly discusses developments in screw and barrel designs from the first reciprocating screw moulding machines to the types in use today. Reproduction of the mathematics of screw design has not been attempted, as it is very adequately covered elsewhere.

The review looks at screw configuration and compares flight depths for differing materials. The use of dual-flight or barrier screws, vented screw/barrel assemblies and mixing heads is also looked at.

Reference is made to the wear occurring on equipment with glass and mineral-filled materials, with some suggestions to reduce this wear.

The development of back-flow valves, adapters and nozzles is reviewed. This covers design details such as mating faces, thread pitches, contact areas and the need to remove any hold-up spots from the complete system. Examples are given showing how inadequate application of design principles can lead to machine and mould corrosion.

1 INTRODUCTION

The development of the reciprocating screw injection moulding machine has taken the length to diameter ratio from 10:1 up to 27:1 for a vented barrel. Screw design configurations are developing with a variety of barrier-type plasticising screws. The old concept of a screw, when processing polyamide, was that the feed section transported the granules to the compression section where all the polymer was changed rapidly to melt. Accordingly, this change was then made in half to one-and-a-half turns. Now it is recognised that the material starts melting in the feed section, with the result that the sharp compression zone is both unnecessary and more difficult to control.

Developments in the design of back-flow valves and the engineering of the barrel accessories have also taken place—all to add to the improvement of polymer handling. However, it is as well to point out that more work needs to be done in these areas, since, if polymers are to reproduce optimum properties in commercially moulded parts, then this area of the machine must be correctly engineered.

In attempting to review developments that have occurred it is inevitable that some may have been missed and very recent work not reported. This may be true of this chapter.

The prime purpose of an injection moulding machine is to plasticise the plastics material so that the plastic mass can then be injected into, and formed by, the mould. Arguably, therefore, the screw and barrel assembly is the most important part of the machine. If this portion of the machine is in any way inefficient, then the efficiency of the whole process is affected. It has been said that the screw and barrel are the heart of the machine.

The concept of screw design has changed since the first reciprocating screw machine was produced in the early 1950s. It is also true that there have been developments in other areas of the barrel assembly which have proved to be almost as important as the most radical changes in screw design configuration.

2 SCREW DESIGN DEVELOPMENTS

2.1 Screw Reciprocation

The mathematics of calculating the basic screw geometry for given plastic melts is well known, and has not changed from the concepts laid down for extruder screw design. An excellent summary of all the formulae and calculations involved is covered by the chapter on extrusion in reference 1. This covers the work needed to deal with extruder screw design (polymer feed, viscous flow, etc.) for calculating the outputs. What is different about the injection machine screw versus the extruder is that the screw reciprocates and does not run constantly.

2.2 Screw Length

The overall length of the screw flights is often used as a reference. However, unless the total length of the screw is effective, in that the rearmost flight land is under the rear edge of the hopper throat, then the length of the screw, divorced from the assembly, is not relevant. The relevant length of screw is that distance from the rear of the hopper throat to the front of the screw, when it is in the fully forward position. It does not include any flights which may be to the rear of the hopper throat, nor does it include any portion of a back flow valve or smear head which may be fitted to the front of the screw.

In considering the developments which have taken place in screw design, one must look at the trends in design which have occurred since the early machines. Initially the screw length to barrel diameter (L/D) ratio was relatively low. Many machines of the mid-1950s to the mid-1960s had L/D ratios as low as 10 to 1. It is much more usual today to see screws with L/D ratios of around 18 or 20 to 1.

The shorter screw was probably due to using the basic machine layout of a piston machine, and merely bolting a reciprocating screw on to the existing bed. Time has shown how much this step restricted the plasticising ability of the machine.

One advantage of increasing the screw L/D ratio is the improved melting and mixing capacity of the screw. To some extent increased power will be needed to turn a longer screw, but much of this will be turned into heat in the polymer melt. The increased length then allows more flights to be devoted to the shear and mixing elements of the screw's functions. It must not be assumed that perfect melt quality is impossible when using a screw with a low L/D ratio. Good quality melt from a short screw can be produced, but not in high quantity. As the output rises from a short screw, then so the quality of melt will reduce.

The longer screw and barrel assemblies, with the attendant lengthening of the machine bed, also allow other features to be added to injection moulding machine screws. Two features now being exploited as options on the injection machine are vented barrels and mixing heads. Both of these have been in use on extruders for many years but have hitherto been unavailable on injection machines. Short screws have certainly been a factor in restricting the development and use of these two features.

2.3 Screw Geometry

Whilst the lengthening of screws and barrels has become a trend in Europe, 20:1 L/D ratios are almost standard US practice. This has given the US moulder more freedom in his choice of screw designs and developments.

Clearly, the loss of five flights on a screw of 15:1 L/D is a restriction of screw design when compared to the 20:1 L/D screw.

The screw has to accommodate materials in various states. This is basic to both extrusion and injection moulding. The raw, unheated solid is first conveyed along the screw. The form of this may be a fine particulate dust, a granular solid, or even an elastomeric strip. This then changes to a mixture of melt and solid. Finally, all the material should be homogeneously molten.

The screw has to be designed to accept the change in volume from solid to melt. This change of volume incorporates effects due to the change of state,

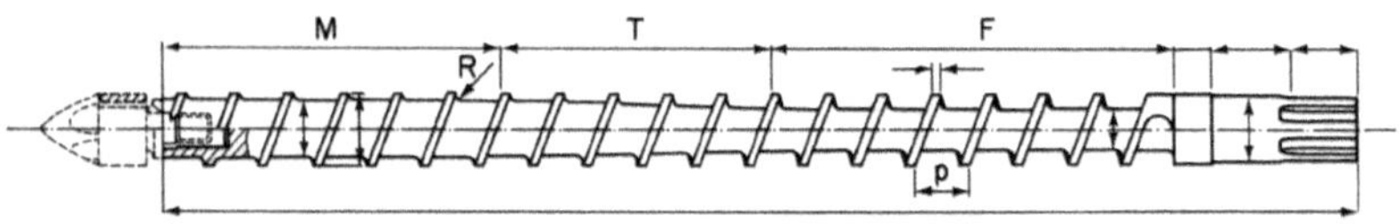

FIG. 1. Metering type screw.

the reduction in bulk density and the expansion of the melt due to temperature rise. Allowance must be made for the greater change in volume which accompanies the change of state of crystalline materials than amorphous materials.

In these circumstances it is clear that the depth of flight at the rear of the screw cannot be the same as that at the front. It must be stressed that this is a consideration of thermoplastic materials and not of thermosetting resins.

Another factor to be considered in designing screw geometry is the susceptibility of the melt to breakdown due to overheating. The method employed in designing screws for PVC is to use a screw which imparts little shear to the material. Heat is added from the cylinder heater bands whilst the material is transported up a constantly reducing flight depth.

The more conventional screw design that is now used is the metering type screw. This screw configuration can be seen in Fig. 1.

The screw consists of three zones; M = metering zone; T = transition zone; F = feed zone. The pitch between the flights is shown as p.

The screw root diameter is constant throughout the metering and feed zones, but not, as already mentioned, at the same diameter. The transition zone is a taper to change smoothly from feed to metering zones. It is sometimes called the compression zone.

These zones are variable and should be dependent upon the type of material being processed. The variations in a given screw diameter and length can be in the length of each zone, as well as the flight depths in the zones. All that has been said so far about extruder screw design also applies to the design of moulding screws.

2.3.1 *The Feed Zone*

The feed zone on a moulding machine is effectively variable in length, due to the reciprocating screw. A typical maximum screw stroke for a 40 mm diameter screw would be 150 mm. This would represent nearly four flights of the screw—20 % of the total length—or 42 % of the feed section length on a 20:1 L/D screw having a feed section of nine flights.

As the term feed section implies, this section of the screw has to deliver sufficient polymer granules to ensure that the screw has adequate melt output. In extruders it is quite critical that the correct amount of melt is constantly being produced at the die. Neither interrupted flow nor surging must occur due to under- or over-supply of melt.

With the moulding machine screw, the totally constant melt supply is not an absolute requirement in all cases, so a certain amount of under-design is currently normal for general purpose moulding screws, when compared to typical equivalent extruder screws.

Both calculation and practical experience in designing extruder screws have led to the conclusion that the channel volume ratio from feed section to metering section should lie between 3:1 and 4:1.[1] This allows for the decrease in volume (2:1) as the granular feed is converted to melt; the maximum volumetric capacity of the metering section; the average efficiency of about 40 % of the solids conveying ability of the screw; and the 20 % variation usually experienced in measurements. However, another factor which may also need consideration is the volume change associated with the latent heat of fusion of crystalline materials (Fig. 2).

There is a trend for leading moulding machine screw designers to raise these channel volume ratios. This was revealed by a survey of general purpose 40 mm ϕ screws available in 1975. Earlier screws had shown channel volume ratios of the order 1·6–1·8:1. The 1975 survey showed that the range was 2·1–2·3:1. This increase in the ratio should help to reduce average screw retraction times by providing adequate feed for the metering zone to pump.

2.3.2 *The Transition Section*

The transition section in a metering-type screw is perhaps hard to define in pure mathematical terms. The major contention is over how many turns the transition should take place. Since both the feed and metering sections are parallel, then it has been proved by experience that an uncomplicated taper from one root diameter to the other is adequate.

It is well to remember that it is in this section of the screw that much of the melting actually occurs. This means that the mechanical change of the screw shape is responsible for the complex physical change from solids to melt. As has been mentioned, this must include any volumetric change

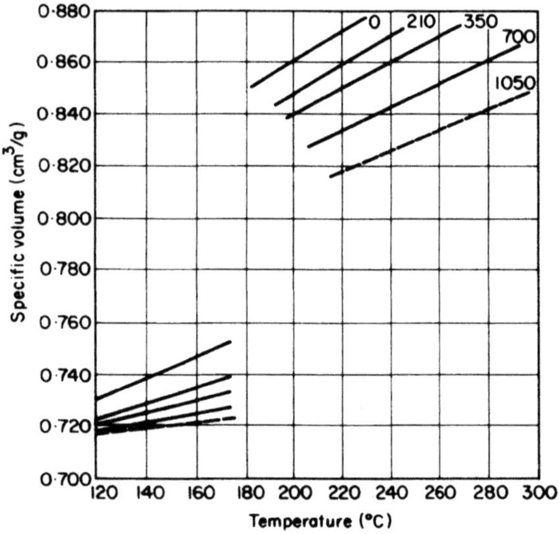

FIG. 2. Pressure–volume–temperature relationship for acetal homopolymer. Pressure: kg/cm^2.

associated with the latent heat of fusion of crystalline materials. Whilst this is almost an instant occurrence, the rapid transition zone of half to one turn of the screw is sometimes difficult to control in a moulding machine. It is possible that the reciprocating motion may have some bearing on this.

In current injection moulding machines, the transition zone is usually between 20 to 40 % of the total length of the screw, which appears to provide flexible enough control under processing conditions.

2.3.3 *The Metering Section*

The metering section for a moulding screw is where extrusion theory and practice almost totally coincide. This is because the theories and calculations for screw melt pumps will apply here.

However, the method of controlling the mixing of the melt as it is pumped differs between moulding machines and extruders. The moulding machine has hydraulic pressure applied to the screw at the drive-end to provide back-pressure. The extruder relies on an adjustable pressure valve in the head or die. The moulding machine sometimes has an accidental back-pressure built into it. This is where the flow of material through the channels of the back flow valve are insufficient to carry all the generated melt from the screw. Further reference will be made to this when the back flow valve is discussed.

In practice the metering section length varies from two turns up to perhaps 40 % of the screw length. The best performance seems to lie around 25 % of the screw length, although the extent would depend upon melt viscosity. The viscosity of the plastic will also affect the metering depth, and, in a properly designed screw, the revolutionary speed should also be defined, otherwise excessive shear of the melt may result.

2.4 Comparison of Screw Performance

The difficulties that result from processing crystalline materials with general purpose screw designs has been reported and noted by Ricour,[2] Canovi[3] and Flieger and Goldman.[4] Whilst the observations have included changes in L/D ratio, channel depths and the length of the three sections, they have been restricted to relatively small diameter screws. This is perhaps because most of the crystalline materials are processed on the smaller equipment. However, the results of the work tie in well with pre-calculated designs using extruder mathematics. It is therefore reasonable to assume that the results can be scaled up for larger equipment.

The results published indicate that finished mouldings exhibit a uniform crystalline structure which can only be obtained with a fully homogeneous melt. In the search for improved part quality and high part performance, the design of the screw is critical and should never be ignored. It can be seen from Fig. 3 how the calculated and recommended metering flight depths for four different material types compare with each other. Clearly, processing problems will result if a screw designed for polymer D is used to process polymer A.

2.5 Dual Flight or Barrier Screws

There are many machines in the USA now fitted with barrier-type screws. These have been designed and developed to improve melt quality and homogeneity combined with high throughput rates. The designs and technology are still under patent restriction, but it is possible to purchase screws from various manufacturers who have licences.

Briefly, the feed channel is split into two, so that there are separate flights

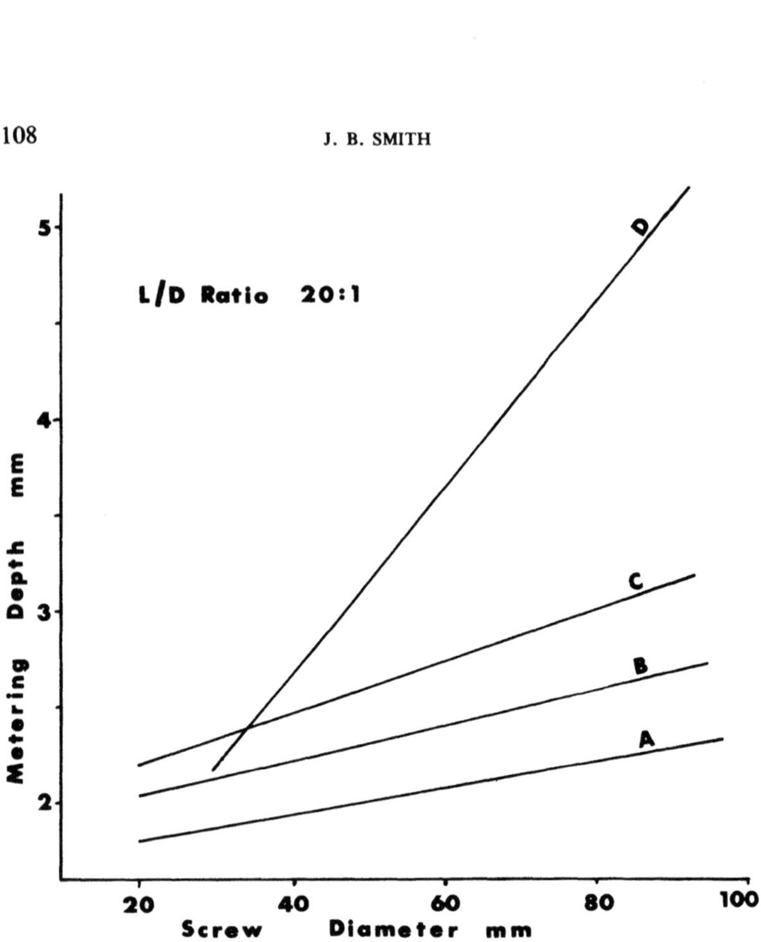

FIG. 3. Metering section depths for four polymer classes.

of differing depths. These two flights are contained between the normal flight, and have a secondary flight separating them with a 0·020 in (0·5 mm) clearance between flight and barrel. The forward flight starts with the original feed section depth and the rearward flight with the depth of the flight land, i.e. 0·020 in. The forward flight gradually reduces its depth over about 11 turns, whilst the secondary flight increases its depth over the same distance, and the 0·020 in land is retained between the two flights. Once the flight tapers have reversed they are merged into a normal metering section.

The effect of this is to gradually change the solids bed from the feed to melt, which is being constantly fed back over the 0·020 in land into the secondary flight. At the end of the taper, the secondary flight is filled with melt and the primary flight ends. The secondary flight then feeds the melt into the normal metering section.

As with normal screws, the dimensions of the flight channels and their length have to be tailored for particular materials to get the best results. Certainly, this is where material specialisation with specific screw designs would be needed and where some of the versatility of the 'general purpose' screw would certainly be lost. Another point which needs making is that these screws should not be employed to process either glass- or mineral-filled materials. The fillers cause excessive wear of the melt weir, and passage of the melt over the weir tends to reduce the filler particle size, thus significantly altering the physical properties of the material.[7] A review of all the currently available barrier screw designs and their variations was published by Chung in 1977.[9]

2.6 Mixing Heads

These devices are well-known in extruders and the original patents have now lapsed so that the original designs are in general use. There are other later designs which seem to be more efficient and which are still the subject of patent cover.

The mixing head or torpedo is added to the front section of a metering screw. This is usually three turns of the screw in length. The torpedo can be smooth or have flights and channels on it. The basic concept of this mixing head is contained in the patent assigned to Dulmage.[10]

The reason for using this type of torpedo head is to improve the mixing effect produced by the screw and to aid dispersion of pigments, etc. It will not improve the melt quality which is solely dependent upon the design of the screw on which it is mounted. It will tend to improve melt uniformity which may result in lower stress in the moulded parts.

Some of the internally moulded stresses in moulded parts are the direct result of poor quality melt. Layers of melt at varying temperatures have been identified as emanating from the screw flights and have not been dispersed or mixed by anything forward of the screw flight. So, if melt of varying temperature in layered streams is introduced into the mould cavity, then the result will be stressed mouldings due to a varying cooling rate. These stresses have been observed in crystalline plastics such as Nylon 66 and identified as a direct cause of weak mouldings.[11]

There have been attempts to market screws which have had mixing heads added to injection machine screws. In one case the screw and barrel was lengthened by three turns to accept the mixing head. In another case, the head was substituted for three of the turns of the metering section.

The former approach was demonstrated to improve the dispersion of even minute additions of pigment when a large shot was injected into a large area.

In the second case, removal of three vital turns of the metering section reduced the quality of the melt from the screw. The result was that the head improved the mixing of the melt being produced by the screw, but the melt was of lower quality. There was little gained from the exercise in this case.

The use of mixing heads in injection moulding machines is in its infancy, and doubtless further developments will occur which will bring about the wider use of these devices.

2.7 Vented Screws/Barrels

The devolatilising screw and barrel assembly is well established for the extrusion of thermoplastics. In the past 5 years or so, many developments have been made to fit these onto injection moulding machines. Again, the basic difference between the extruder and the moulding machine has caused some design differences.

The incentive to use vented equipment arises from a need to remove either volatiles naturally occurring in polymer melts, e.g. methyl methacrylate in acrylics, or, more commonly, water vapour. Before the introduction of vented barrels, the only way to remove these was to dry the materials thoroughly prior to processing. The very variable efficiency of the drying systems used has often meant that moisture extraction was not always complete.

Figure 4 shows, schematically, the arrangement for a typical vented barrel assembly. To ensure that the volatiles can be removed, it is necessary

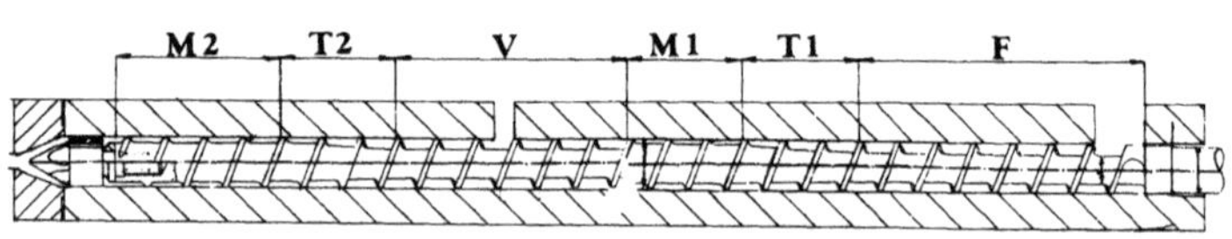

FIG. 4. Vented screw and barrel.

to plasticise all the solids first. It follows, therefore, that a normally designed screw section is the first requirement. This is indicated by F (feed section); T1 (first transition section) and M1 (first metering section). The compressed melt with the volatiles must then be allowed to decompress. This is done by increasing the flight depth such that the melt only partially fills the flight channel. The volatiles are then exhausted through the vent port. The venting section V must be long enough to allow the vent always to be over the parallel screw root, no matter how much of the reciprocating stroke is being used. The devolatilised melt is then recompressed in T2 (second transition section) and remixed in M2 (second metering section).

There can be a problem in ensuring that only molten plastic is fed to the vent section. Clearly, if a screw design is developed for one particular material it is comparatively simple to ensure that only melt reaches the venting region. One way of achieving this for a screw designed to accept a variety of materials is to use a weir. This is rather like a smear head except that it is built into the screw. It only allows melt to pass into the venting area, thus facilitating the removal of volatiles.

Another problem is that of ensuring that the venting section of the screw remains only partially full. Again this can be solved by screw design for individual material cases. However, the method which is now becoming popular, particularly on larger screw units, is to employ some means of starve-feeding the plasticising screw. This is being done by the use either of a weigh feed device, or of a continuous vibration feed, with adjustable flow rates.

It clearly is not just as simple as cutting a hole in the barrel wall. Certain problems in shape and location of vent port, relative to the screw centre line, are outlined by Simms and Maycock [6] Another point which is also made by them is that the minimum L/D ratio needed for a vented screw processing 66 nylon is 27:1. The reasoning behind this can easily be deduced from Fig. 4, and it is difficult to dispute the logic of their argument.

Vented barrels can prove to be a benefit in day-to-day processing of moisture-sensitive polymers such as acrylics, polycarbonates, ABS, and, to some extent, nylon. Where the moisture level is below 1 %, then there is a marked advantage. However, process cycle times can rise unacceptably if polymers having significantly higher levels of moisture are attempted. The vented equipment will, of course, still process materials which are essentially dry. Under these circumstances it may be necessary to block the vent port to exclude atmospheric oxygen. This could otherwise produce oxidisation of the plastic melt, with the attendant reduction both in appearance and in physical properties. Another way of dealing with this is the application of an inert gas blanket. There have been reports from Japan that indicate this method has been tried, using dual ports to 'wash' the oxygen and moisture from the vented area.

At this stage opinion is still divided over the success and limitations of vented barrel usage in injection moulding. No doubt as more experience is gained in their use, then the benefits or otherwise will become clear.

3 INJECTION MOULDING MACHINE BARRELS

3.1 Barrel Wear

The bulk of injection machine barrels made in Europe are produced with nitriding of the inner surface to protect it against wear. The alternative to this is to use a bi-metallic liner in the barrel. This may be cast *in situ* or, as in the USA, a separate liner may be inserted and shrunk in place. This difference in the method of manufacture may explain why 95 % of injection machines sold in the USA have bi-metallic cylinders as a standard feature. The European percentage is almost the complete reverse on original equipment specification.

The need for a bi-metallic liner is because of the increasing use of glass- and mineral-filled polymers and, to a lesser extent, polymers having highly corrosive decomposition products.

Various sources have indicated that the improvement in wear resistance found when processing glass-filled materials using a bi-metallic liner is a factor of 10:1. This is without considering treatment to the screw and back flow valves. Simms and Maycock[5] demonstrated the differences between nitriding, chrome plating and ion-nitriding both on barrels and screws, etc.

However, when considering improved wear performance of bi-metallic lined barrels, one should also seriously consider treatment of the screw. Various treatments have been tried on the screw, which range from adding 'Stellite' tips to the screw flights, ion-nitriding the whole screw, to the use of the Union Carbide WT-1 detonation gun. This latter development, which is a means of applying a special wear-resistant coating to a screw, is fairly recent, but early results have been spectacularly successful when processing glass-filled and mineral-filled nylons.

The bi-metallic coating of check-ring sleeves and seats is also used, but excellent results can be obtained by using high chrome steels (10–12 %), e.g. Röchling RCC Supra or Boehler special KNL hardened to Rockwell C 60–62. A very small clearance between check ring and barrel (*c.* 0·01 mm) is also important in this area.

3.2 Prime European Suppliers of Bi-metallic Barrels and Screws

Trade Mark	*Supplier*
'Bernex'	Bernex, CH 4600 Olten, Switzerland.
'Brux'	Brooks (Oldbury) Ltd, Brades Road, Oldbury, Warley, West Midlands B69 2DL, UK.
'Reiloy'	Metall GmbH, 521 Troisdorf, West Germany.

'Xaloy'	Xaloy S.A., Hauts-Sarts Industrial Park, Route de Hermée, Herstal-Lez-Liège, Belgium.
'WT-1 Gun'	Union Carbide Europe S.A., 5, rue Pedro-Meylan, CH-1211 Geneva 17, Switzerland.

4 INJECTION UNIT ACCESSORIES

4.1 Definition and General Considerations

It is difficult to consider an injection machine screw and barrel without a back flow valve (or smear head for PVC), an adapter (or end-cap) or a nozzle. However, these parts are often changed for alternative designs for use either on different materials or on moulds. As a result, they may be considered 'accessories'. None the less, they are frequently the cause of many of the problems associated with processing polymers.

As a preamble to dealing with these parts the following line of reasoning is applicable:

1. No polymer is infinitely stable at, or above, its melt temperature.
2. All polymers will therefore degrade at or above their melt temperatures as a function of a time period.
3. Many of the resulting degradation products of polymers have a corrosive effect upon the materials of construction of the injection machine barrel assembly and the mould.
4. The polymer *melt* should *all* be transferred to the mould on a continuing basis.
5. Thus, there should be no crevices, holes or hold-up spots where any melt would remain for any period of time, however these might have arisen.

From this it should be clear that not only is the basic design of the machine components critically important, but equally their maintenance and very careful assembly are also critically important. Whatever good work may have been done in ensuring a good design of screw, etc., it can all be negated by poor attention to detail at the melt end of the assembly.

Many of the difficulties arise from basic design in the areas where the 'accessories' are joined to their neighbouring components. It is these 'mating faces', that will be considered first, as they are common to the joints of the back flow valve to the screw, the adapter (end-cap) to the barrel, and the nozzle to the adapter. It is a matter of 'getting it together', to put it in modern parlance.

4.2 Machine Construction

4.2.1 *Joint Faces*

A basic engineering requirement for joining round or cylindrical parts together, is that parallel diameters only should be employed. Further, if streamlined flow is important, these diameters should be registered and finished to very close tolerances.

Three areas of difficulty arise from this:

1. The basic design may be faulty.
2. The design may be correct, but the quality of manufacture may be poor, or
3. The parts may have been damaged either in assembly or whilst being maintained.

Any of these situations will cause corrosion and ultimate wear. The reason for the occurrence of faulty joints does not matter; the end-result is always the same, i.e. streamlined flow will not be achieved.

4.2.2 *Inappropriate Tolerances*

In many instances, a designer's objective is to make manufacture 'easy'. The logic is that, if manufacture is 'easy', it is therefore less costly. One of the techniques used with this in mind is to widen the tolerance band. The problem arises when the tolerance band is widened at a point where the chemical engineer would deem a close tolerance to be mandatory. In Fig. 5 the effect of not matching cylinders to precise tolerances can be seen. In the example on the left, where one cylinder is larger than the other, material will sit in the hold-up spots (a and b), degrade and cause corrosion. The example on the right shows what can happen if the screw root is not ground concentrically with the valve fixing hole; this again causes a hold-up spot (c). For joint faces anywhere in the melt path, the maximum tolerance on both mating parts should be no greater than $\pm 0{\cdot}01$ mm ($\pm 0{\cdot}0004$ in) in diameter.

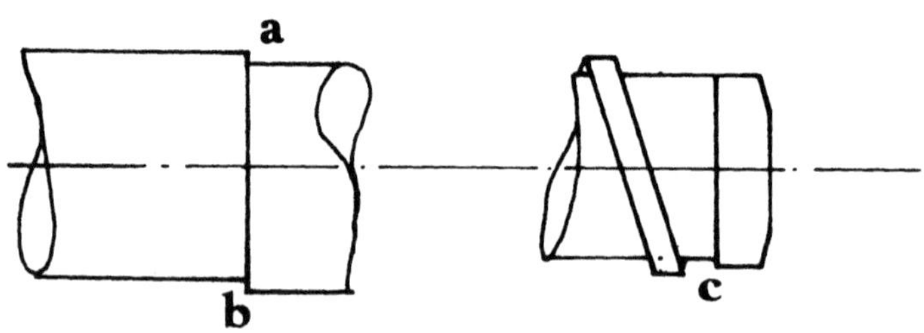

FIG. 5. Examples of non-matching cylindrical parts.

4.2.3 *Taper Joints*

In one sense, the objective of 'easy' manufacturing leads on from tolerances to taper joints. These may be taken to be either in 1, 2 or 3 position in Fig. 6. Example 4 shows basically how the joint ought to be made, but this is not necessarily trouble-free. All have been used, or are in current production use, in injection machines as sold ex-manufacturer's works.

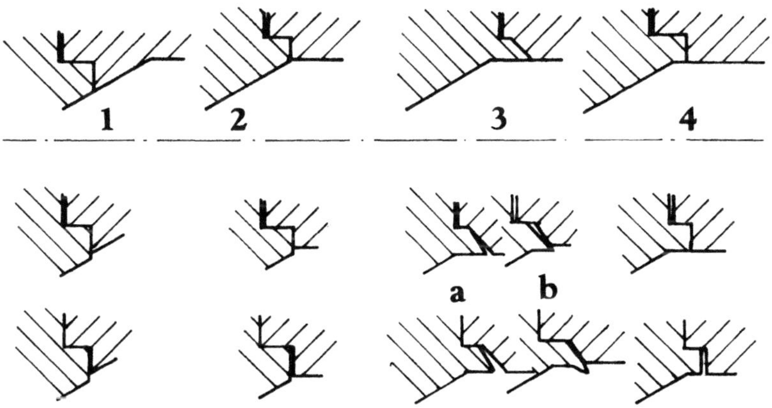

FIG. 6. Examples of tapered joints and resultant problems.

Example 1. This shows the classic taper-to-taper joint face. Even if one assumes that it is made perfectly in the manufacturing stage, which is a difficult engineering job, then the first time that one face is 'cleaned up', the faces no longer match and a hold-up spot results. Further cleaning will remove the face clearance and produce a larger hold-up spot and the associated corrosion.

Example 2. This is essentially similar to Example 1, with the exception that one face is cylindrical. The end result is a hold-up spot.

Example 3. The basic 45° angle face joint appears on many machines, despite a determined effort to have such designs adjusted. There are two possible ways that this can develop into a problem. Example 3a shows what happens when the edge is not 45° but a little more; 3b is when the situation is reversed. Example 3a produces an easily identifiable hold-up spot and 3b is what usually occurs when the fitter is ensuring a good firm joint, ably assisted by a 6 ft length of hydraulic pipe attached to the spanner. The weak front edge is compressed and curls up. This creates a hold-up spot which is rapidly corroded.

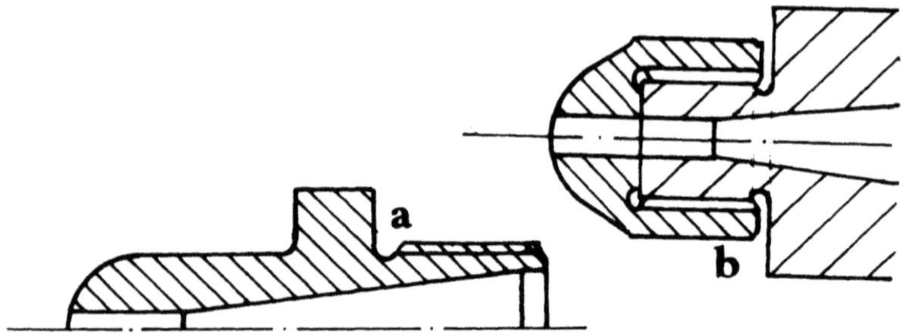

FIG. 7. Examples of nozzles having no thread register.

Example 4. Essentially this is how joints should be made. If it is necessary to clean the joint it can easily be done and all will be well. However, a careful eye should be kept on the thrust clearance so that the final sketch in that sequence does not occur.

4.2.4 *Lack of Register*

A fault that continually recurs, is that of failing to provide adequate location for two mating parts. The most common example is where a screw thread is used as a register. The thread should only be used to provide a tensile load; it can never be accurate enough to provide good location.

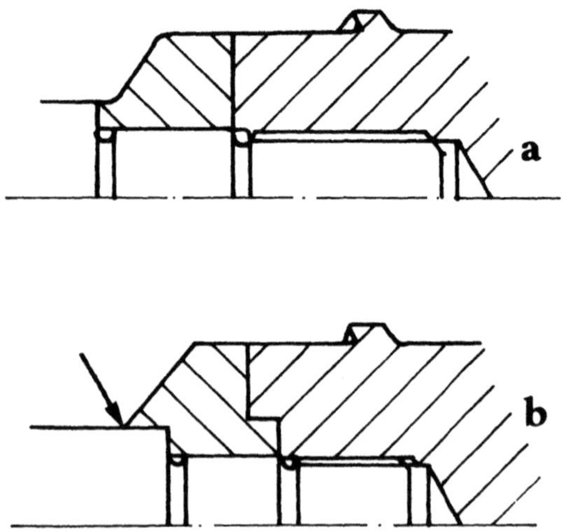

FIG. 8. Examples of poorly registered back flow valve threads.

A place where the thread is improperly used is in the nozzle-to-adapter joint, if it is designed without a closely toleranced register. Two examples are shown in Fig. 7. Example (a) is the most common, and misalignment of the orifice is frequently found. Example (b) is less common, but it is still an unregistered thread. This has the added disadvantage that there is no direct temperature control on the nozzle tip.

Examples of no or limited register are shown in Figs. 8(a) and (b). In Fig. 8(a) a well-designed valve seat is located on the tip, but the tip is not located into the screw.

Figure 8(b) shows minimal location, by using a larger spigot on the valve seat as the location. This is not enough, and adequate location is usually at least as long as the thread diameter. Another problem here (arrowed) is the sharp edge. This will quickly allow material to get between the edge and tip, with the resulting hold-up spot and ultimate corrosion.

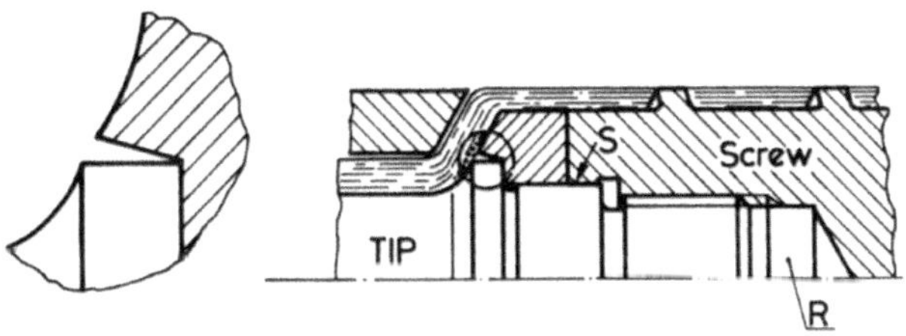

FIG. 9. Result of poor register on back flow valve seat.

Figure 9 shows another design of valve tip and seat. The locating shoulder S is not big enough to prevent the end of the tip being moved off-centre on start-up. This leads to the opening of the seat as in the scrap or enlarged section and the creation of the hold-up spot.

Figure 10 shows how the location is improved and the tip retaining the seat on matched and located diameters. This design has proved to be trouble-free in many machines over the last 6 years.

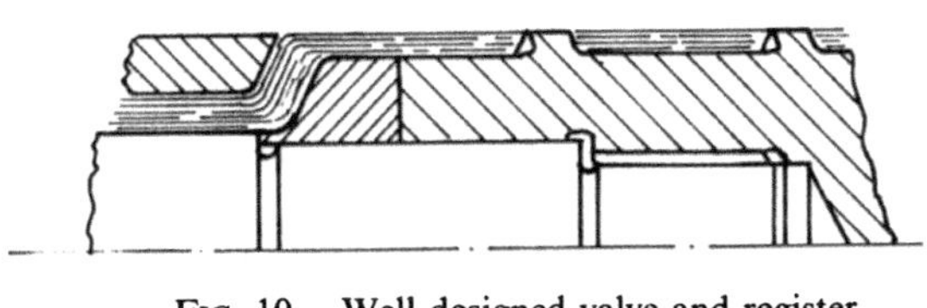

FIG. 10. Well designed valve and register.

4.2.5 *Joint Face Widths*

Where two mating faces are to be joined by means of a thread, the face width of the contact area can sometimes cause some problems.

If the face width be too narrow, then the possible overtightening by fitters can lead to the edge being crushed, and result in a hold-up spot.

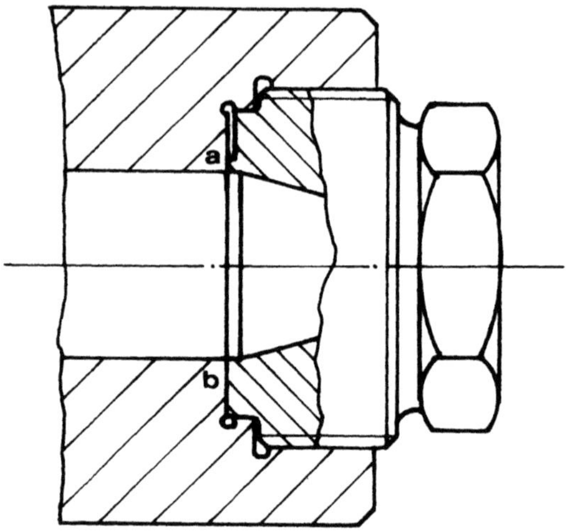

FIG. 11. Examples of seat widths on threaded adaptors.

In Fig. 11, *a* indicates a too narrow land which may well be crushed by overtightening the thread, whilst *b* has a face that is too wide and thus it may not be possible to generate sufficient force to close this face. This might well occur if some molten material were accidentally trapped between the faces. The face width should be between 5 to 8 mm. One feature which is sometimes used is shown in the enlarged scrap section (Fig. 12) which is to back-taper the male face by half a degree. This ensures a close-fitting interface joint.

4.2.6 *Manufacturing Quality*

Machine manufacturers do not produce all the parts for their machines themselves. Clearly some specialist equipment has to be bought in. It is also usual to find smaller repetitive parts purchased from outside suppliers. When this business is being placed, a price is negotiated. This is then a

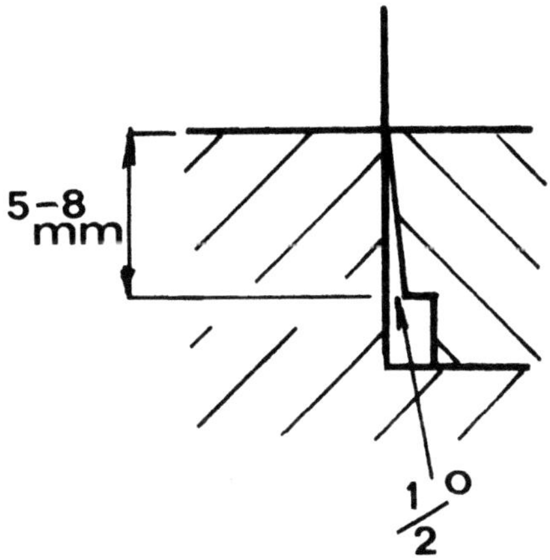

FIG. 12. Enlarged view of seat face.

major factor, and unless the quality control side is strictly maintained, then trouble can be caused to the eventual machine purchaser.

In Fig. 13(a), the original joint as designed can be seen, and Fig. 13(b) shows how one part has been designated 'sharp edge'. This is vital. Only by having two square sharp edges which then match perfectly when registered together, can hold-up spots be avoided. If either or both edges are radiused then the hold-up spot is present and machine corrosion is then assured. These are illustrated in Figs. 13(c) and (d).

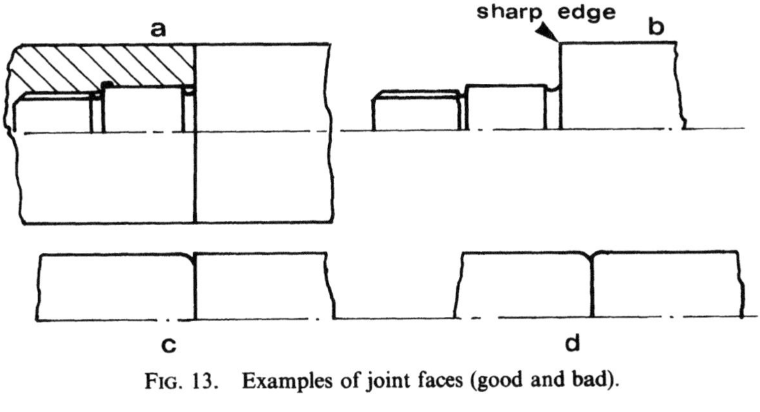

FIG. 13. Examples of joint faces (good and bad).

4.2.7 *Damage*
It is possible to damage machine parts mechanically either in assembly or during maintenance operations. This may not be intentionally caused, but the effect is none the less bad.

Figure 14 shows how the edge of one part may become chipped. If the part is re-assembled with this chipped edge, the result will be a material hold-up spot.

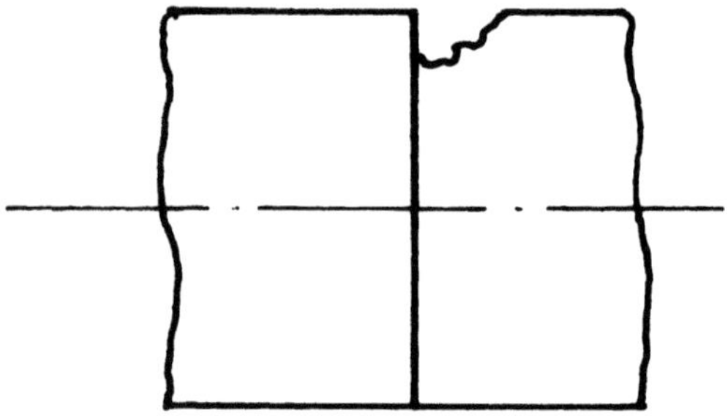

FIG. 14. Damaged edge.

4.3 Back Flow Valve
This is fitted to the front of the screw, to allow injection pressure to be exerted on the melt and into the mould cavity. The major exception to this is when processing rigid PVC, and generally a smear head is then used.

There are very many designs of back flow valve. In the history of the reciprocating screw moulding machine, it is probable that lack of attention to this valve's design has caused more process problems than perhaps any other machine feature.

The prime function of the valve is to prevent the melt from being forced backwards over the screw when injection pressure is applied. It must also open and allow all the melt generated by the screw to pass freely through it.

The designer of the back flow valve must have these points in mind: (a) no flow restriction; (b) no hold-up spots; (c) a perfect seal; (d) low wear rate of component parts.

4.3.1 *Flow Restriction*
The output of melt from the screw should be governed by the screw design and by outside control such as applied back pressure. It must not be controlled by the back flow valve.

Should there be a restriction within the valve, then this will result in an extended screw refill time, possible overheating of the melt and increased wear of the screw/barrel assembly, because of the increased time taken to refill.

In short the valve, when open, should be capable of allowing the passage of all the melt generated by the screw without creating any artificial back pressure.

4.3.2 *Hold-up Spots*
There should be no built in hold-up spots in the back flow valve which must be designed and manufactured in such a way that all the previously mentioned problems of mating joint faces and unregistered location are faithfully avoided.

4.3.3 *Perfect Seal*
Should the back flow valve fail to seal repeatedly on every cycle, then the result will be a wide variation in moulded part quality. Failure to seal may well result in shrinkage variation, poor mechanical properties, voids and surface defects.

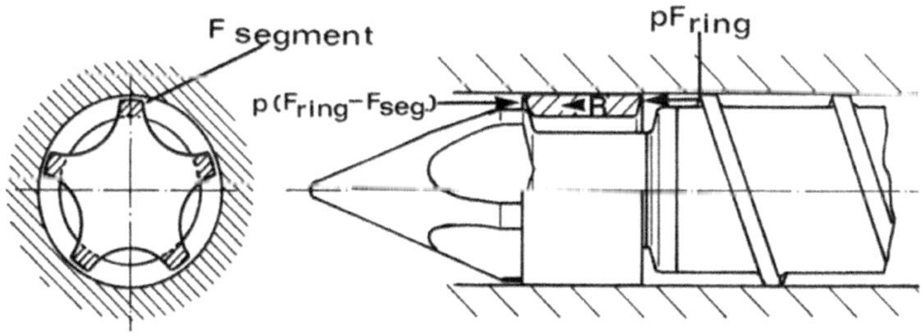

FIG. 15. Forces applying to ring check valve.

Figure 15 demonstrates a ring check valve and the forces acting on it to effect the closing seal force.

The ring of this valve is sealed when the formula condition is reached:

$$p(F_{ring} - F_{segment} + R) < pF_{ring}$$

R is the frictional force between the ring and the barrel wall.

4.3.4 *Wear Rate of Component Parts*
The fact that moving parts will wear must be accepted. The rate of wear, in some respects, is governed by the design, materials of construction and the

treatment of the part's surfaces. Once the basic notion is accepted that parts will wear and will then need replacing, then the designer may proceed.

In some cases the designer has tried to ensure that no wear is to take place and, further, that no maintenance is likely to take place either. Under these circumstances, he is ignoring the behaviour of the plastic melt and the customer's real requirements. The result is usually a penalty of increased screw retraction time and an increase of barrel wear.

Canovi[8] points out that the most economic solution is to try to ensure that the simplest, cheapest part wears first. This then should be replaced 'more often', so that the basic machine performance is maintained. The user should be conditioned to check the wear on a valve every 6 months to 1 year, and, if necessary, change the worn part.

In the case of the sliding ring back flow valve, the wear should be directed at the ring. One way of doing this is to make sure that it is softer than the other parts it contacts. If the head and seat be Rockwell C 52 and the ring Rockwell C 44 this will suffice. One steel that has provided good results on a general basis is Boehler MPD or WMD extra as an alternative grade.

Some adjustments to steels and hardnesses may be necessary if the special requirements for glass- and mineral-filled materials are necessary.

4.3.5 *Types of Back Flow Valve*

Essentially two main types of back flow valve have been made, the ball check valve and the sliding ring check valve.

The ball valve designs consist of a fixed head attached to the screw. This head has channels bored in it with a restriction built in. A ball or balls are allowed to float in the restricted area of the seat, so that, when injection pressure is applied, the ball(s) close(s) the hole orifice(s), preventing material passing back up and over the screw flights.

One example of a ball valve is shown in Fig. 16. This is a multi-ball design and many machines have been built using a similar layout. This type of valve exhibits built in hold-up spots, notably between the inlet holes and

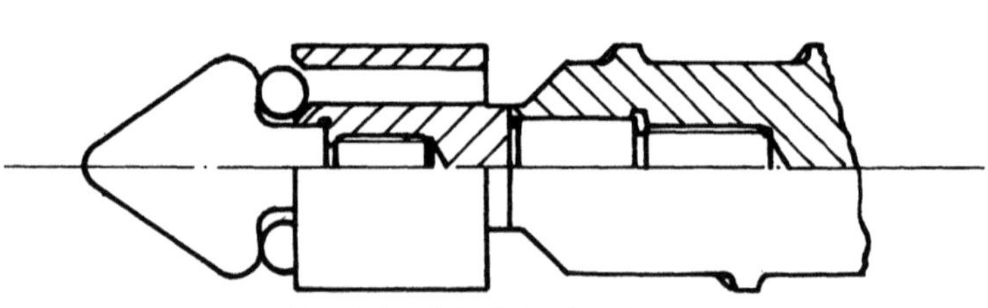

FIG. 16. A ball check valve.

behind the balls. In some cases the balls tended to flatten and become lozenge-shaped. They did not then seat correctly and the valve would no longer hold the injection pressure—its prime function. Figure 17 shows how one such valve has cracked round the ball seats (A) and how some corrosion has started on the register (C) and thread (D), due to the chipped edges (B).

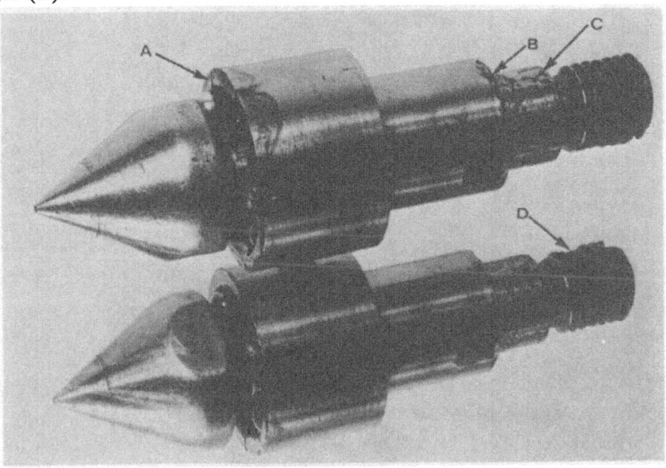

FIG. 17. A worn ball check valve.

Clearly, the problems that have occurred with this valve have resulted from not following the requirements which have been laid down as principles.

There are other designs of ball valve which only use a single ball centrally located in the head. The problems with all ball valve designs are (a) not being able to completely streamline the flow so there are no hold-up spots, and (b) higher manufacturing cost because of their complexity.

The sliding ring check valve is in much wider use. It has the basic benefit of being uncomplicated in principle. The result should be to produce the perfect valve each time. Unfortunately, this is not the case, as the four attributes necessary for a valve are not always kept in mind by the designer. It may not be his fault directly, as he has to satisfy other interests in addition to designing a good valve. These may be that the valve should have as universal a use as possible, in that it should fit as many screws as possible and be suited for all plastics. There is also a natural inclination that it should differ in design from valves designed by other machine makers. In

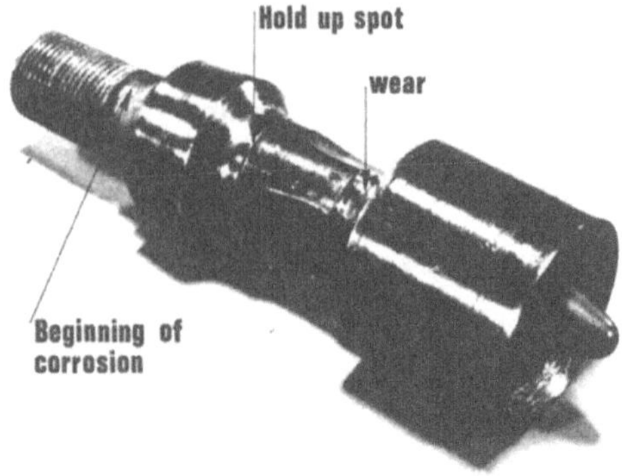

FIG. 18. Corroded and worn ring check valve.

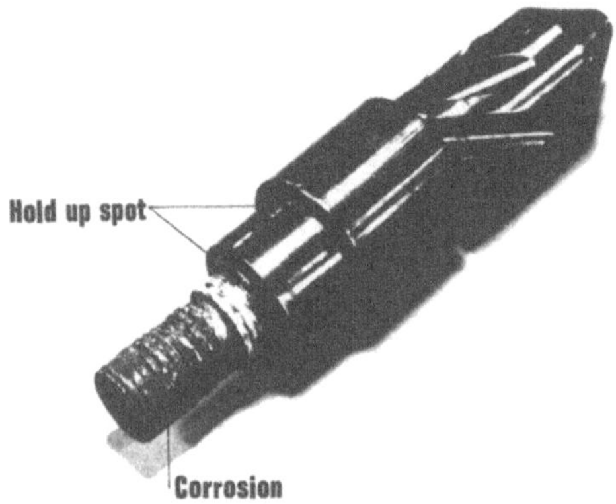

FIG. 19. Ring check valve with hold-up spot and subsequent corrosion.

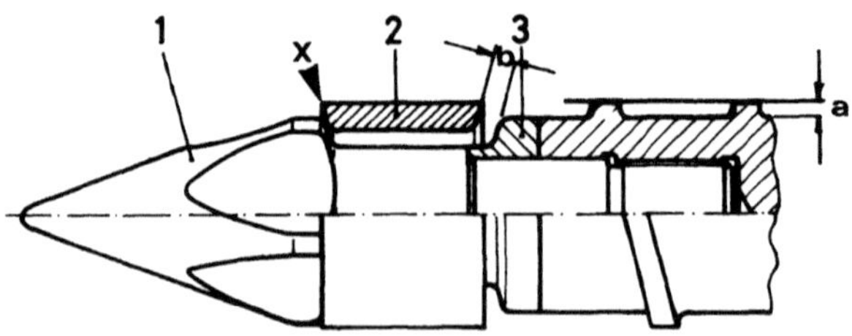

FIG. 20. Flow path proportions for ring check valve.

considering these facts, it is easy to see how valves are made which have some deficiencies. Figures 18 and 19 demonstrate some of the possible results.

This type of valve really needs three pieces to be successful. It has been produced with two—by using the screw face as the seat—and as many as five pieces.

Figure 20 shows the layout of a carefully designed back flow valve using the three necessary pieces. The design incorporates all the requirements needed for a successful valve.

The dimensions are such that the valve opening, dimension (b), should always be 1 mm more than the flight depth of the screw, dimension (a). Further, the clearance between the head stem and the inside diameter of the ring should also be greater than (b). The five large scallops (located between points 1 and x) are sufficient in area to ensure that no back pressure is created at that point.

The joint faces are mated on sharp edges to close tolerances ($\pm 0{\cdot}01$ mm on the mating diameters), and they are located by a good register. The length of this should be at least one diameter of the fixing thread inside the screw.

All other edges in the melt path are carefully radiused, so that there is no tendency to hang up and excessive melt shear does not occur.

Figure 21 shows the effect of correctly mating the valve to the screw. It should be noted that sufficient radius was not applied to the front edge of the valve seat when this was made. The whole assembly was designed as a replacement for the ball valve shown in Fig. 17.

Finally, the front edge of the back flow valve, indicated with an 'X' in Fig. 20, should be no further from the end of the barrel than 4 mm, when the screw is in the fully forward position. This helps to ensure that material is always swept through the system and does not tend to stagnate against the heated barrel wall.

The dimensional properties and tolerances for the valve are given by Canovi.[8] The valves using these proportions have been proved to work on many different makes of machine, with excellent results. In addition to proving successful with acetal and nylon, they have also been shown to be successful with polycarbonate, polystyrene and several other polymers, but not with rigid PVC.

A back flow valve may sometimes be used to reduce the effect of screw design deficiencies. This is usually a speedy expedient made as an alternative to avoid fitting a completely new screw. Figure 22 shows the way this has been done. The length of the valve seat is increased so that a short

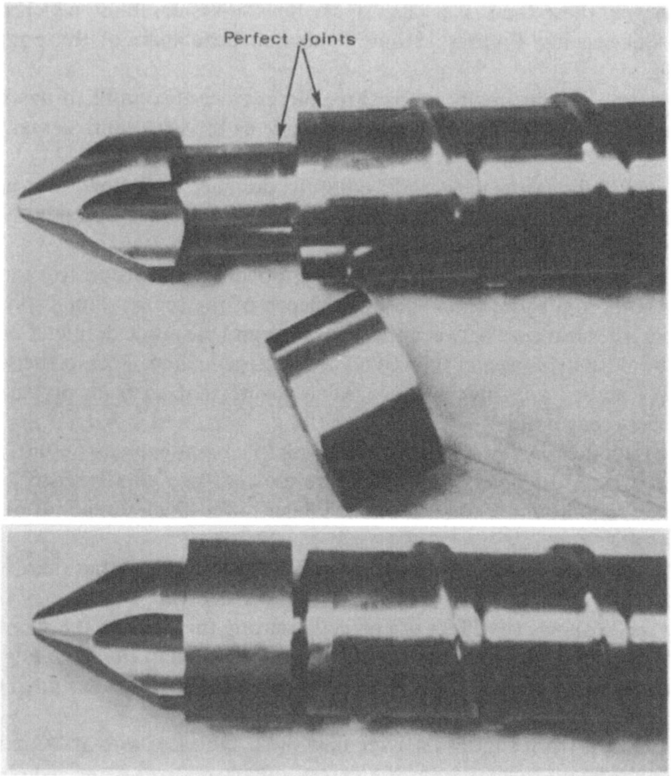

FIG. 21. Correct mating of valve tip to screw.

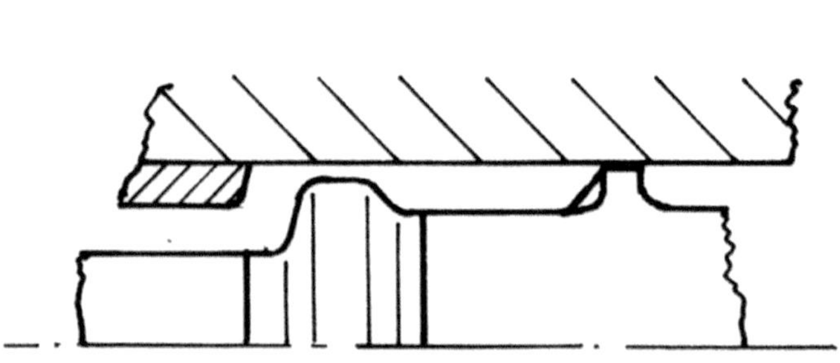

FIG. 22. A smear wier to partially overcome the effects of an overdeep metering section.

weir can be incorporated at the end of the screw. This has the effect of increasing back pressure in the screw and of improving the plasticising ability. It is usually resorted to on a screw having too large a metering section flight depth.

4.4 Adapters

Much of what has been written regarding joint faces and hold-up spots will apply to adapters. These parts are sometimes not used in smaller machines, as the nozzle is directly threaded into the barrel end, thus serving two functions, both as nozzle and adapter. As barrel size increases, this solution becomes more difficult.

However, the adapter not only serves the purpose of reducing the barrel diameter down to an appropriate size suitable for a nozzle, but it also acts as an insulating piece. A nozzle screwed into the barrel sometimes is very difficult to control thermally, particularly when a precise nozzle tip temperature is needed. The use of an adapter helps to reduce the thermal overload on the relatively smaller mass of the nozzle compared to that of the barrel.

A practical feature of the moulding shop operation is that, to some extent, nozzles are made specially for individual tools. This results in a nozzle being changed at the same time as the tool. Frequent nozzle changes are very likely to damage the seating area, with the result that hold-up spots and corrosion will then occur. If this occurs on a machine without an adapter, the damage will be on the seat in the barrel. This may well prove to be both expensive and difficult to rectify. The use of an adapter will transfer the vulnerable seat to an easily replaceable part, which solves a potential maintenance problem. Canovi[8] offers a suggestion for an adapter which shows the relationship of the back flow valve tip to the adapter and nozzle (Fig. 23). It is stressed that this solution is not acceptable for rigid PVC.

The adapter will, in general, have no additional heat applied to it, as it should not be necessary to apply further heat to the molten plastic inside. This aids the more delicate nozzle tip temperature control.

It must be noted that adapters are also bolted onto the end of the cylinder. This system has been used with many machines and has been shown to be workable. However, much care needs to be taken by the machine designer to ensure that delicate thermal balance in this area is maintained. Some machines have exhibited a degree of over-kill on the temperature control systems applied, and this has resulted in extreme difficulties in controlling the temperature.

4.5 Nozzle

The purpose of the nozzle is to direct the molten material from the cylinder into the mould. It is necessary to avoid undue flow resistance and to maintain the plasticised material at a uniform temperature.

4.5.1 *Nozzle Temperature Control*

The nozzle temperature control is basically to prevent the material chilling off and blocking the orifice; it is not for adding heat to the molten plastic.

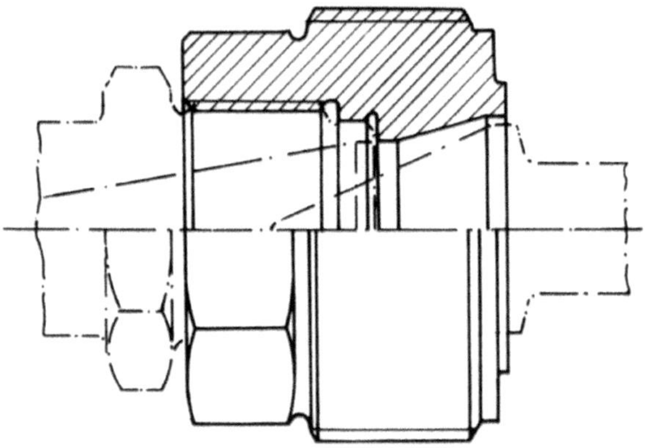

FIG. 23. Relationship between screw tip and adaptor (not for PVC).

The most important area for temperature control is the tip which is in contact with the sprue bush of the mould. The further away from the tip that the thermocouple and any heater band are, then the more difficult will the temperature control be.

There are many machines offered with no heater band applied to the nozzle. This means that control and maintenance of temperature depend upon a remote thermocouple and heater. This cannot be looked on as good practice. Further, the temperature is sometimes controlled by means of a time-based energy controller. It is on or off for a period of time dependent upon the time set by means of a varying length bi-metallic strip. This again, is not good control practice in this very sensitive area, even if the material being processed is the most thermally stable melt available.

The best results are usually obtained when the nozzle has its own independently controlled and sensed heater band. The control should be the type that results in the heater band always being on, using a proportion of the available power. A Variac or its electronic equivalent have proved to be the most suitable for this.

4.5.2 *Contact Area Shape*

With the requirements for temperature control in mind, the other features of nozzle shape in the contact area are based upon mechanical needs.

Both of the views as to whether the nozzle should centre on the sprue bush or not have strong advocates.

The flat face nozzle view is held because of the logic that the mould should not be responsible for aligning a machine barrel and screw. It is usually accepted that this will make the tool setter's job more difficult.

The opposing view, that there should be alignment between nozzle and sprue bush, is based upon the need to ensure that the nozzle orifice aligns exactly with the sprue centre. This is usually done by unequal radii. The radius of the sprue bush seat is larger than the radius of the nozzle tip, which ensures that the nozzle tip and sprue align exactly and maximum contact pressure is generated immediately adjacent to the orifice. The nozzle tip radius should be between 14 to 16 mm and the sprue bush radius 18 to 20 mm, with a difference of 2 mm respectively.

Should the radius on the nozzle be too small, the effect will be to decrease the contact area. The danger is that the orifice edge will peen over, with repeated impacting, and this will restrict flow through the nozzle. The same effect will arise if a conically shaped tip is used due to the 'feather edge' at the orifice.

4.5.3 *Fixing Threads*

It is common for machine designers to use threads to join the nozzle to the adapter or end-cap. In many cases, the adapter is itself also threaded into the barrel. This system is most generally used on machines up to around 500 tons' clamping force.

When the thread pitch is selected, care should be taken to ensure that it is neither too fine nor too coarse. Should the thread be too fine, then there willl be a high risk of crossed threads when fitting and of thread seizure due to oxidation and corrosion. If the thread pitch be too coarse, then there is a risk of the thread loosening due to vibration and impact during the machine operation. Experience has shown that the best thread pitch for barrels of 30–80 mm diameter is around 2·5–3 mm. This applies both to nozzle–adapter and the adapter–barrel joints.

4.5.4 *Inside Form Shape*

Many shapes have been tried inside nozzles, for many different reasons. The variations have proved endless as designers have striven to overcome control problems of both machine and melt.

To some extent the need for the very small orifice on the thermal seal nozzle and the various shut-off nozzles on machines has gone. This is the

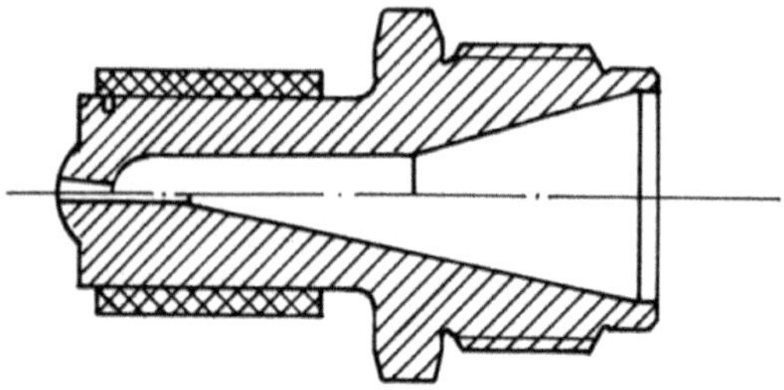

FIG. 24. Two internal configurations of nozzle (top: amorphous plastics, bottom: crystalline plastics).

direct result of the machine feature of screw decompression, which relieves the residual pressure in the cylinder following screw retraction.

The nozzles now in general use either have a general taper from adapter to a parallel orifice, or have a small reverse taper orifice fed from a larger parallel bore. Figure 24 shows examples of both these forms, and also summarises the other points of nozzle design.

The thermocouple is placed towards the nozzle tip to ensure good control. Also, the heater band is well forward so that heat input is directed to the tip, which is the most likely region to freeze and block.

REFERENCES

1. BERNHARDT, E. C. (1959). *Processing of Thermoplastic Materials*, Reinhold Publishing Corp., New York.
2. RICOUR, B. (1972). *Plastics and Polymers*, August, 217.
3. CANOVI, P. N. (1975). Some developments in internal design of screws and barrels, Paper given at PRI Conf., *Advances in Injection Moulding*, 11th–12th November.
4. FLIEGER, H. D. and GOLDMANN, K. (1975). *Kunststoffe*, **65**(144), 193.
5. SIMMS, C. J. and MAYCOCK, D. F. (1977). *Plastics and Rubber Weekly*, 25th February, 17.
6. SIMMS, C. J. and MAYCOCK, D. F. (1973). *Some Comments on the Use of Vented*

Injection Cylinders for Processing 'Wet' Type 66 Nylon, Imperial Chemical Industries Plastics Division.
7. FILBERT, W. C., JR (1969). Reinforced 66 nylon–molding variables vs fiber length vs physical properties, *SPE Journal*, **25**(1), 69.
8. CANOVI, P. N. (1975). Design considerations for plastifying unit, Du Pont de Nemours International, S.A., TR G 50-7508.
9. CHUNG, C. (1977). Plastic engineering, *SPE Journal*, February, 36–7.
10. US Patent 2,453,088, assigned to Dulmage, 2nd November, 1948.
11. BOEHME, E. (1974). *32nd Annual Technical Conference, Safety—Professionalism—Ethics*, 13th–16th May, 1974, San Francisco.

Chapter 5

MATERIALS MODIFICATION IN INJECTION MOULDING

R. C. McGREGOR
Technical Adviser, Marketing Services Department, ICI Plastics Division, Welwyn Garden City, UK

SUMMARY

The paper takes as its theme the modification of existing polymers to fit new and existing applications, since there is unlikely to be any large-scale introduction of entirely new polymeric materials. The need for modification of polymers with particular regard to their engineering properties is discussed, followed by a survey of other means of modifying existing materials including copolymerisation, filling with various fillers, and foaming.

1 INTRODUCTION

It is over a century since Alexander Parkes told the Royal Society of Arts, 'Another important improvement in the manufacture of Parkesine is the employment of camphor, which exercises an advantageous influence on the dissolved pyroxyline, and renders it possible to make sheets, etc., with greater facility and more uniform texture, as it controls the contractile properties of the dissolved pyroxyline; camphor is used in varying proportions according to requirements, from 2 to 20%'. By this modification to the otherwise intractable material he helped celluloid on its way to its unassailable position as the world's first man-made plastic. From this humble modification has grown the industry with its multitude of polymers and polymer variations which are known today.

The days of trial and error in research polymerisation of all available monomers, and the rapid introduction of the new polymers, have, however, gone. Chemists now understand the basic reactions involved in polymer chemistry and can make knowledgeable predictions as to those likely to have economically advantageous qualities. Those materials relying on cheap and readily available feedstocks have also been fully investigated. In the future new materials will undoubtedly be introduced from time to time, to extend the range of properties at present offered by existing materials. These are likely to be expensive and of limited application when compared with the established large-tonnage polystyrenes, PVCs, and polyolefins.

Therefore, if injection moulding is to progress from a material as well as a mechanical standpoint, this progress must lie in the modification of existing polymers to improve their performance in existing applications and to develop entirely new uses for these modified materials.

2 THE NEED FOR POLYMER MODIFICATION

The need for modification of polymers arises from the inherent limitations of the base materials. The modifications are an attempt to overcome these limitations by providing a balance of properties, which are otherwise not available, tailored to the end application, without sacrificing ease or economy of processing.

The main limitations of non-modified polymers are as follows.

(a) *Strength and surface hardness.* Plastics are not as strong as traditional materials, particularly metals, and the surface hardness is inferior to that of conventional substances such as glass. The hardest thermoplastics surface is displayed by the acrylics and this is similar in performance to aluminium. There are at present no modifiers to improve the surface hardness other than a coating of silicone materials, for example. However, the incorporation of graphite in nylon will give a measure of improvement by increasing crystallinity in the surface layer of the polymer.

(b) *Flexural Modulus.* The flexural modulus of plastics materials is an order of magnitude less than that of traditional high stiffness engineering and construction materials, as shown in Fig. 1. However, by employing sections giving maximum stiffness, and/or by incorporating fillers, notably fibrous fillers such as glass, in the material, it is possible to minimise this disadvantage (as shown in Fig. 2).

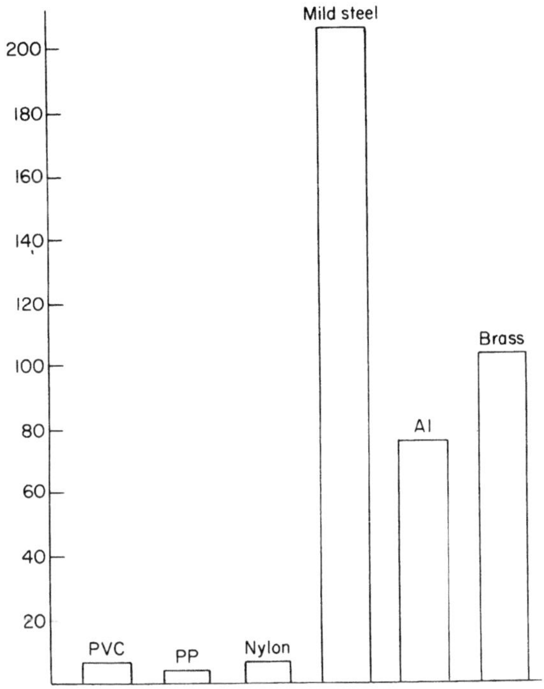

FIG. 1. Comparison of the flexural modulus of various materials (GN/m^2).

(c) *Temperature resistance*. The upper limit for operation of components injection-moulded from commercially available materials is set at around 200 °C by polyethersulphone. The most commonly used large-tonnage plastics such as polypropylene and acrylonitrile butadiene styrene (ABS) are limited to around 120 °C. The maximum temperature of operation cannot be increased to any great extent by the use of additives or modifiers though such properties as stiffness may be preserved to a higher temperature within the ultimate working limit of the material. Low temperature flexibility of, for example, polypropylene can be markedly increased by copolymerising with ethylene, which itself (as polyethylene) withstands cryogenic temperatures without difficulty.

(d) *UV resistance and outdoor weathering*. In the unmodified form there is a wide range of resistance to weathering, typified by the acrylics on the one hand, which withstand many years of weathering in tropical conditions

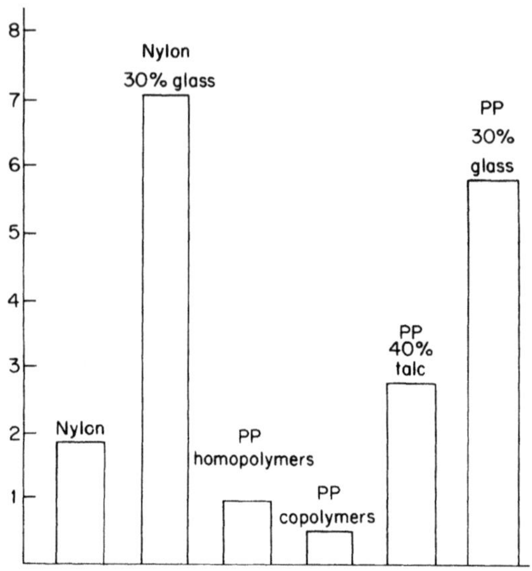

FIG. 2. Comparison of the flexural modulus of modified polymers (GN/m^2).

with negligible deterioration, to polystyrene and polypropylene on the other, which without modification deteriorate, to an unacceptable level of strength, within very few months in strong sunlight. It is common practice for most plastics materials to be offered in grades containing ultra-violet stabilisers.

(e) *Flammability*. All plastics materials which are polymers of hydrocarbons will burn to a greater or lesser extent. Cellulose nitrate will burn almost explosively, rigid PVC will not support combustion in the absence of a flame, whilst a piece of polyethylene will burn like a wax candle. Grades are now available with modified burning characteristics obtained by the incorporation of additives in the original polymer. These materials have to be treated with caution since the additives, in many instances, merely increase the temperature at which ignition will take place and often result in the production of toxic fumes and smoke on subsequent burning of the component or structure.

(f) *Thermal expansion*. Thermal expansion of plastics varies between 5 and 10 times that of metals and is normally allowed for in the design of the component. However, by the incorporation of suitable fillers it is possible to reduce the expansion within certain limits.

(g) *Electrostatic charges.* Since polymers are poor conductors of electricity it follows that they will be prone to a build-up of static charges. This can vary from being a simple nuisance when, for example, an article of acrylic furniture gathers dust, to a severe hazard when, for example, film is being reeled at high speed in an explosive atmosphere. Additives are available to minimise the effect of static build-up, usually working by the incorporation of chemical reagents which attract a minute film of water to the surface of the component, thereby providing a conducting path for any static electricity.

3 MODIFICATIONS

Virtually all polymers are modified to some extent by the incorporation of additives such as heat stabilisers to facilitate processing, UV absorbers to lengthen service life, colour (for appearance), and so on. However, these additives, though constantly being added to in number and improved in efficiency, are not within the scope of this chapter.

We are here concerned with the more fundamental modifications. These modified materials may also incorporate the type of additive mentioned earlier for enhanced properties. It should be stressed that often when a material is modified to improve a particular property—for example, stiffness—there may be a consequent deterioration in another attribute such as impact resistance. What is needed is the optimum balance of properties for any particular application. Bearing this in mind the various methods of modification will now be examined.

3.1 Copolymers

Copolymers are a means of combining the attributes of two or more individual materials to give a final product displaying the best qualities of each individual while suppressing the undesirable qualities. They are manufactured by polymerising mixtures of the various monomers as opposed to a post-polymerisation compounding procedure. The properties of the copolymer depend on the characteristics of the particular monomers used, the proportion of each monomer, and the type of copolymerisation achieved, i.e. random, block or graft. One of the commonest examples of a graft copolymer in use today is ABS which is available in a number of different grades and which, through usage and utility, has come to be regarded as a polymer in its own right rather than as a modification of another material.

3.1.1 *Olefinic Copolymers*

It has been found possible to copolymerise propylene and ethylene in almost any proportions to give a range of materials varying from rigid thermoplastics to thermoplastic elastomers. For example, propylene is often copolymerised with ethylene (up to about 15% ethylene) and the copolymers differ basically from homopolymers by their much greater shock resistance under cold conditions, although at higher temperatures

TABLE 1
COMPARISON OF IMPACT STRENGTHS OF POLYPROPYLENE HOMOPOLYMER AND COPOLYMER
(ft lb/in notch)

Temp. (°*C*)	*Homopolymer*	*Copolymer*
23	1·5	2·5
0	0·8	1·7
−20	0·5	1·2

there is a slight loss of rigidity and hardness. Table 1 gives a comparison of the impact strengths of a homopolymer compared with a copolymer at various temperatures.

EVA copolymers are copolymers of ethylene and vinyl acetate. Their characteristics range from semi-rigid, translucent types (somewhat similar to low density polyethylene) to rubbery, crystal-clear materials having properties typical, in some respects, of plasticised polyvinyl chloride and certain rubbers. The dominant factors controlling the properties of EVA copolymers are the molecular weight and the percentage of vinyl acetate combined in the copolymer. It is normal to incorporate vinyl acetate up to 18% by weight and the range of molecular weights of material suitable for moulding is expressed in terms of their melt flow indices which range from 2 to 40. Table 2 shows the effect of vinyl acetate content and melt flow index on the balance of properties of the material.

Certain of the EVA copolymers and the ethylene/propylene copolymers belong to a fairly new class of material called thermoplastic elastomers. These will be considered in the next section.

3.1.2 *Thermoplastic Elastomers and Polymer Blends*

Thermoplastic elastomers or rubbers are based on two-phase copolymers and exhibit the properties of vulcanised rubber while having the processing ease of thermoplastics. A well-known example of such materials is SBS

TABLE 2
PROPERTIES OF EVA COPOLYMERS

Copolymer (ICI grade numbers)	Melt flow index[a] (g/10 min)	Vinyl acetate content (%)	Nominal density[b] (kg/m^3)	Tensile strength at break[c] (MN/m^2)	Elongation at break[d] (%)	Stiffness modulus[e] (MN/m^2)	Softening point[f] (°C)	Shore hardness[g]	
								A scale	D scale
501	2	7·5	926	16·2	700	156	83	91	48
554	4	12·5	931	16·2	700	94	72	90	44
566	17	14	933	15·2	700	71	64	90	41
571	8	15	934	16·2	700	67	66	90	40
538	2	18	937	17·2	750	47	64	89	35
540	10	18	937	16·2	700	45	59	87	35
546	40	18	937	12·3	650	45	—	85	35

Test methods:
[a] BS 2782:105C; ASTM D1238-73; ISO R 1133.
[b] (D) of BS 3412:1966; ISO 1872.
[c,d] BS 2782:301F; ASTM D1248-74; ISO R 527.
[e] 100 s, modulus at 0·2% strain.
[f] BS 2782:102D; ASTM D1525-75; ISO 306.
[g] ASTM D 2240-75; ISO R 868.

(styrene–butadiene–styrene) which is a thermoplastic copolymer in which the styrene acts as a thermoplastic material holding together the rubbery butadiene. This type of block copolymer displays self-reinforcement due to the dual-phase nature of the system. Polyurethane thermoplastic rubbers have been known for many years and this class of material has now been expanded by the introduction of the SBS elastomers mentioned earlier and, more recently—following the introduction of EVA copolymers—by the olefinic and polyester condensate types.

Olefinic copolymers based on fully saturated copolymers of ethylene and propylene have varying block structures. Their processing and service temperatures are somewhat higher than those of the SBS types. They do not exhibit clear melting points as such, but soften progressively before reaching viscosities normally associated with processing operations.

The polyester condensate type of thermoplastic rubber, whilst exhibiting true elastomeric behaviour and being capable of processing on conventional injection moulding machines is generally somewhat harder than the majority of rubber compounds, having a minimum hardness of around 90° shore A.

Whilst polymer blends as such are not new—PVC/nitrile rubber blends have been used for many years—it is within the domain of thermoplastic rubbers that polymer blends show most promise for the future. Compounding thermoplastic rubbers with other polymers, giving a multiphase end product, modifies the ratio of the elastomeric to thermoplastic types of property available. For example, high impact polystyrene may be used in combination with thermoplastic rubber of the SBS type for applications such as shoe soles. Another example of this type of compounding is Du Pont's Nordel which is based on ethylene–propylene–diene monomers (EPDM). This may be blended with other polymers such as polypropylene to produce a wide range of different materials. Such blends may be produced by melt mixing.

Processing thermoplastic elastomers. Because thermoplastic elastomers can easily be tailored to give specific properties, in the majority of cases they may be used as supplied with little further modification. Fillers may be added to modify tensile and hardness properties and in many cases are used to reduce cost. Thermoplastic elastomers may be injection moulded without difficulty. However, screw machines are recommended because of their ability to produce a melt with good temperature uniformity and lower stress levels in the final product. Because of the large range of materials available, it is not possible to generalise on processing conditions, and in each case the manufacturer's recommendations should be adhered to.

Thermoplastic elastomers have to date been used in numerous rubber replacement applications but it is the automotive industry which would seem to be the most promising future market, with car bumpers probably accounting for a major share.

3.2 Fillers

3.2.1 *Glass*

The major advantageous effect of adding glass fibre to plastics is to increase the stiffness and the heat deflection temperature (see Table 3). The glass

TABLE 3[1]

COMPARISON OF GLASS FIBRE-FILLED AND UNFILLED NYLON 66

Property	*Units*	*ASTM test*	*Glass-filled*	*Unfilled*
Specific gravity	—	D792	1·38	1·14
Tensile strength	10^3 lbf in^{-2}	D638	23	11·5
	(MN/m^2)		160	80
Elongation at break	%	D638	3–5	80–100
Flexural modulus	10^5 lbf in^{-2}	D790	11	4·3
	(MN/m^2)		8 000	3 000
24 h water absorption at saturation	%	D570	5·6	8·9
Coefficient of linear expansion	in/in °F	D696	1.65×10^{-5}	5.6×10^{-5}
	(cm/cm °C)		2.8×10^{-5}	9.9×10^{-5}
Deflection temperature under load				
66 psi	°C	D648	254	200
264 psi	°C	D648	245	75

fibres are normally sized with materials such as silanes which provide lubrication between the glass fibres. This sizing can provide good interfacial bonding or wetting between the polymer and the glass fibres and, where this is the case—as with nylon—there is a high degree of reinforcement. In addition to the increase in stiffness, this results in a big increase in tensile, flexural and compressive strength, the increase depending mainly on the glass fibre content. The glass fibres normally incorporated for injection moulding materials have lengths between 0·125 mm and 0·5 mm (0·005–0·02 in).

The addition of glass beads gives higher rigidity and compressive strength to the unreinforced material but does not provide the same reinforcement as do fibrous fillers. Mouldings produced from bead-filled

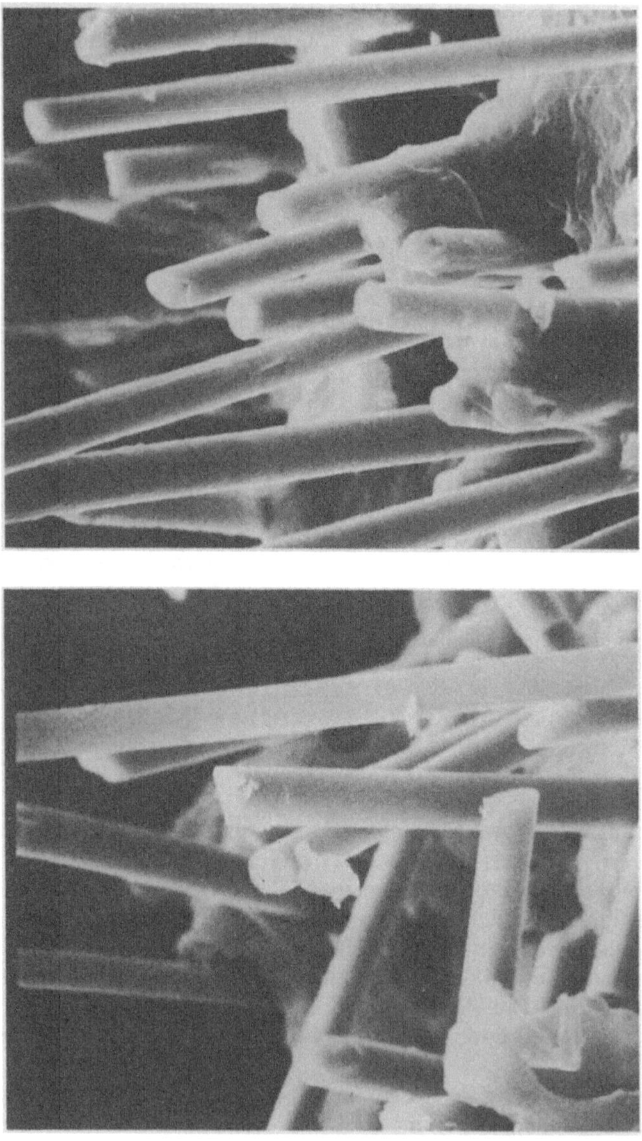

FIG. 3. Electron micrograph of a glass-filled polyolefin fracture surface. (a) without coupling agent, (b) with coupling agent.

material are less prone to warping than those reinforced with glass fibres because of a reduction of the anisotropic effect.

With olefinic materials the interfacial bonding is less strong, and although materials are available which use normal sized glass (known as 'glass-filled grades') it is also possible to employ special coupling agents which greatly increase the strength of the bond between the glass fibres and the polymer, as shown in Figs. 3(a) and (b). These materials are known as 'glass-reinforced grades'. Without this coupling effect the fibres would tend to slip out of engagement under load and the greater rigidity would not be matched by commensurate improvement in load-bearing capacity or creep resistance. A fully coupled system enhances rigidity, tensile strength and creep resistance and brings all three properties into effective balance.

Since glass reinforcement increases the properties of rigidity, tensile strength, creep and heat resistance, fatigue strength, environmental stress cracking resistance and chemical resistance, its use is significant in applications in which high temperatures, vibration, high loading and difficult working environments are involved. Its major use, therefore, is in the improvement of performance of engineering thermoplastics. Typical applications for such modified materials are under-bonnet components in motor cars, mechanical components in domestic appliances and other industrial uses where the good chemical and electrical properties of the polymer are retained along with enhanced mechanical performance. Some typical applications are illustrated in Figs. 4 and 5.

Glass-filled materials—processing. The surface of glass-reinforced mouldings is usually less glossy than that of normal grades, particularly in large areas and in dark colours where some 'flecking' from fibres just under the surface may be visible. The effect is minimised by the appropriate moulding techniques and is often indiscernible, especially in light-coloured mouldings. Sink marking is generally less likely to be apparent than is the case with normal materials and this does permit rather more latitude in the variation of section thickness.

The actual moulding conditions will of course be governed by the particular material, machine and component, but in general the injection pressure for glass-reinforced grades is of the same order as that required for unfilled polymers. However, whereas high injection rates give a more glossy surface finish, slow injection rates generally produce mouldings with less fibre orientation and consequently greater freedom from distortion.

Weld lines arising from converging flows can weaken the material at the interface because of fibre orientation, the effect becoming more serious as flow lengths increase. Gate positions and flow lengths should be selected to

FIG. 4. **Rear and front view of car light housings made from glass-reinforced** (25 %) polypropylene.

avoid the creation of weld lines or to minimise flow lengths. Thus, multiple gates may be preferable for moulds where weld lines occur at the extremities, in order to shorten the flow lengths.

Gate sizes similar to those used for unfilled materials are recommended. Runners should be as short and direct as possible and the best sections for good pressure transmission with minimum usage of material are circular or trapezoidal. The cross-sectional area is dependent on the length of runner but typically diameters of 6–7 mm (0·24–0·28 in) are employed.

The shrinkage is significantly lower than that of normal polymers, with a lower value in the direction of maximum fibre orientation. As the most marked orientation of glass fibres is likely to occur in narrow sections, these are more likely to have shrinkage at the lower levels. Mouldings of large areas will generally possess less marked orientation, depending to some

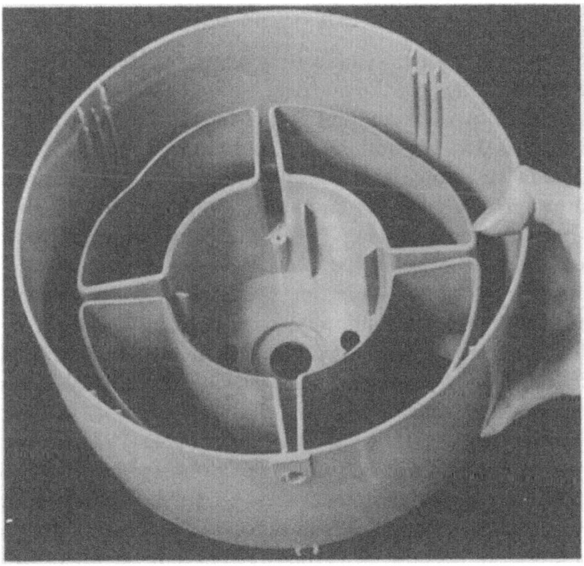

FIG. 5. A fan housing-unit moulded from glass-reinforced nylon 66.

extent on wall thickness. The predominant orientation in many mouldings is not necessarily in the line of flow; first, orientation in the surface layers may be counteracted by orientation normal to this direction in the centre of the section, and secondly, the longer fibres in mouldings tend to orient across the line of flow under some conditions. When precision mouldings are required, because it is not always easy to predict flow directions and orientation, the usual technique, to allow for shrinkage variation, is to design core forms slightly oversized and cavity forms slightly undersized. Adjustments can then be made by removing metal from the mould after initial proving trials. Once mould sizes have been established, the lower overall shrinkage of glass-reinforced grades means that moulding sizes are less affected by variations in processing conditions and, hence, somewhat closer tolerances can be held.

Glass-filled materials are more abrasive in their nature than unfilled polymers and as such will increase the wear on screws and barrels. It is common practice to increase the time before the extra wear becomes excessive by using hardened screws and liners, for example bimetallic liners and ion-nitrided screws.

3.2.2 *Talc*

Compounding talc with polymers is a relatively inexpensive means of extending the usefulness of the material by increasing the rigidity and the dimensional stability of mouldings and reducing shrinkage in the mould. These advantages over the unfilled polymer are, however, accompanied by a lower impact resistance and a reduced resistance to long-term heat-ageing, although both can be minimised by a suitable choice of base polymer and stabilising system. Where there are stringent performance requirements, it may be preferable to use glass-reinforced grades for rigidity and high temperature resistance or to design for the use of unfilled copolymer grades where a high impact resistance is specified.

Talc-filled materials such as polypropylene are used in applications where extra rigidity is needed, coupled with low raw materials costs and the high chemical resistance that is characteristic of polypropylene. Applications include domestic appliance components, where it is currently used for washing machine agitators, pump covers and impellers and where the excellent environmental stress cracking resistance in repeated contact with hot soap and detergent solution is necessary. A typical application is illustrated in Fig. 6. Because of its rigidity at elevated temperatures and its cost advantages, talc-filled polypropylene is often preferred to other thermoplastics for car heater boxes, heater box covers and impeller motor supports. Its cost advantage is used to the full in non-returnable containers for tomatoes, citrus and soft fruits, green vegetables, etc. Further reductions in cost are possible because the high rigidity of the talc-filled material allows the use of thinner wall sections than are feasible with unfilled polypropylene.

Talc-filled materials processing. The natural colour of talc-filled grades is a light grey or buff, and therefore the brightness of certain colours is reduced. This base colour can also have a detrimental effect on pastel shades. Although specially selected talc is normally used, variations in the colour of the filler are largely outside the control of the polymer manufacturer or compounder and therefore there may be some batch-to-

FIG. 6. Tumble drier cabinet made from talc-filled polypropylene.

batch variation in the colour of talc-filled materials. This can pose problems where close colour control is required.

Talc is a hygroscopic material and packages containing it should be sealed during storage. It is also recommended that talc-filled blends be stored under dry conditions.

The mechanical properties of talc-filled materials are influenced by the proportion of talc present and by the base polymer used. Table 4 lists the physical properties of the principal talc-filled grades of Propathene polypropylene, where it can be seen that an increase in rigidity is achieved at some slight sacrifice of tensile and drop impact strength. For example, a 40 % talc-filled medium impact copolymer has a flexural modulus over twice as great as that of an unfilled homopolymer and a drop impact strength at 23 °C which is marginally less than that of an unfilled homopolymer at the same temperature.

TABLE 4

TYPICAL PROPERTIES OF TALC-FILLED GRADES OF 'PROPATHENE'®

			'Propathene' homopolymer grades			'Propathene' copolymer grades		
Powders			522/21	545/21	522/24	601/21	—	601/24
Granules			22/21	45/21	22/24	101/21	202/21	101/24
Talc content (% by weight)			20	20	40	20	20	40
Property[a]	Test method	Unit						
Melt flow index (MFI)	BS 2782: 105C ASTM D1238-70							
Powder	190 °C/10 kg		12·0	40·0	12·0	12·0	—	12·0
Granules	230 °C/2·16 kg		2·0	5·0	2·0	2·0	4·0	2·0
Density		g/cm³	1·03	1·03	1·22	1·03	1·03	1·22
Tensile strength at break	ASTM D638-64T (2 in/min)	lb/in²	5 300	5 300	5 000	4 100	3 750	4 050
		kg/cm²	373	373	350	290	264	285
		MN/m²	36·5	36·5	34·5	28·8	25·8	28·1
Flexural modulus	ASTM D790-66	lb/in²	490 000	490 000	621 000	460 000	350 000	510 000
		kg/cm²	34 400	34 400	43 700	32 000	24 600	36 000
		GN/m²	3·38	3·38	4·28	3·14	2·41	3·55
Izod impact strength at:	BS 2782-306A							
23 °C (73 °F)		ft lb/in	0·62	0·50	0·55	0·90	1·00	0·76
0 °C (32 °F)		notch	0·43	0·41	0·42	0·60	0·67	0·62
−20 °C (−4 °F)			0·42	0·40	0·43	0·60	0·67	0·50
Rockwell hardness	ASTM D785-65	R Scale	105	104	100	98	87	94
Softening temperature	BS 2782-102D	°C	150	150	150	147	148	147
	ISO/R306	°F	302	302	302	297	298	297
Heat distortion temperature	ASTM D648-56							
66 lb/in² (4·6 kg/cm²) (455 kN/m²)		°C	120	120	130	113	112	123
		°F	248	248	266	235	234	253
264 lb/in² (18·6 kg/cm²) (1 820 kN/m²)		°C	74	74	88	67	65	77
		°F	165	165	190	153	149	171
Flammability (These results relate only to the behaviour of the test specimens under the particular conditions of the test. They are not intended as a means of assessing the particular fire hazard of the material in use.)	ASTM D635-74 (3 mm thickness)	Burning rate cm/min	2·3	2·6	2·1	2·3	3·0	2·0
	BS 2782-508A (AMD 1524) (1·5 mm thickness)	Burning rate mm/min	40	45	35	40	45	35
	FMVSS302 (2 mm thickness)	Burning rate mm/min	35	38	38	38	35	38
	(3 mm thickness)	mm/min	23	25	23	23	25	23
Coefficient of thermal expansion at 20 °C (68 °F)		Per °C	$0·8 \times 10^{-4}$	$0·8 \times 10^{-4}$	$0·55 \times 10^{-4}$	$0·8 \times 10^{-4}$	$0·8 \times 10^{-4}$	$0·55 \times 10^{-4}$
		Per °F	$0·45 \times 10^{-4}$	$0·45 \times 10^{-4}$	$0·31 \times 10^{-4}$	$0·45 \times 10^{-4}$	$0·45 \times 10^{-4}$	$0·31 \times 10^{-4}$
Mould shrinkage		%	0·8–1·8	0·8–1·8	0·9–1·3	0·8–1·8	0·8–1·8	0·9–1·3

[a] The properties of plastics are dependent on many factors, e.g. temperature and length of time over which the stress is applied, and therefore most of the results in this table should not be used for

The material retains a high resistance to chemical attack typical of the unfilled material, but oxidation resistance at elevated temperatures is reduced by the addition of talc, and mouldings may become embrittled after long exposure to high temperatures. This reduction in oxidation resistance is caused by the preferential adsorption of stabilisers by the talc.

For the injection moulding of talc-filled materials, in-line reciprocating screw injection moulding machines are recommended, because they can provide efficient plasticisation. The moulding conditions required will be governed by the particular machine and moulding; however, a melt temperature of around 230 °C is typical for talc-filled polypropylene. Because of the presence of the filler, weld lines arising from converging flows can weaken the material at the interface. Gate position and flow lengths should be selected to avoid creation of weld lines and to minimise flow lengths. Thus multiple gates may be preferable for moulds where weld lines are causing problems, and in order to shorten flow lengths. Gate sizes similar to those employed for unfilled material are recommended, while runners should be as short and direct as possible.

When talc-filled materials are moulded the surface is typically less glossy than that of unfilled grades. The degree of gloss is dependent on the level of filler being used, but a uniformly silky finish can be achieved. Sink marking is generally less likely to be apparent than in moulded unfilled materials and this does permit rather more latitude in the variation of section thickness.

3.2.3 *Asbestos*

Asbestos may be used as a filler to maintain mechanical properties, particularly stiffness, to a temperature nearer the ultimate working temperature of the material. When using asbestos-filled materials it is important to realise that the normal standards of safe working conditions laid down for the use of asbestos must be observed. It is common to incorporate asbestos in a fire-retardant grade of the base polymer.

3.2.4 *Miscellaneous Fillers*

For improved surface lubrication of mouldings, graphite and molybdenum disulphide have been used for many years either on their own or in combination with glass. In addition to improving the slip properties, improvement in surface hardness is apparent when used in crystalline polymers, resulting from an increase in crystallinity at the polymer surface. Typically a glass- or graphite-filled material produces improved mechanical, thermal and frictional properties. (See Table 5.)

A wide variety of other mineral-filled polymers have been marketed from

TABLE 5

COMPARISON OF THE COEFFICIENTS OF FRICTION OF VARIOUS GRADES OF NYLON 66

(Dry, i.e. no lubrication; sliding speed: 80 cm/min)

'Maranyl' grade		*Coefficient of friction*				*Comments*
		Against steel	*Against brass*	*Against A100*	*Against A190*	
A100	Static	0·34	0·40	0·42	0·51	Unreinforced
	Dynamic	0·23	0·33	0·24	0·33	
A190	Static	0·29	0·36	0·43	0·41	Glass fibre reinforced
	Dynamic	0·21	0·33	0·22	0·23	
A198	Static	0·33	0·49	0·44	0·39	Glass fibre and graphite reinforced
	Dynamic	0·20	0·42	0·24	0·27	

time to time; the fillers include chalk and various forms of silicates. These materials are primarily intended to cheapen the product and in their formulation attempts are made to obtain a balance between reduced performance and economic use of the polymers. Because of the high density of these materials they are often not as cost-effective on a volume basis as would appear at first sight and the justification of their use is therefore sometimes somewhat dubious. It is often worth considering whether it would not be preferable to cheapen the component by optimising the design and using a less expensive base polymer.

Organic titanate coupling systems may also be used with mineral fillers to improve the flow and impact behaviour and increase the rigidity.

4 FOAMED MATERIALS

There has been a large increase in recent years in the use of foamed materials, particularly structural foam used for load-bearing applications, such as those shown in Fig. 7. Foaming represents a convenient method of obtaining higher rigidity per given weight and therefore per given cost of material. Materials which may be foamed for injection moulding include polystyrene, ABS, polypropylene, polyethylene, nylon, PVC and polycarbonate, in addition to the commonly available polyurethanes. Whilst it is possible, in foam moulding, to produce a wide range of densities depending upon the actual moulding conditions and grade of material used, it is normal when considering items to be used in engineering

FIG. 7. Structural foam pallet made from polypropylene.

applications to achieve density reductions of around 30%. Other advantages of the process include the ability to produce large area thick section mouldings and stress-free mouldings, since the process relies on expansion of the blowing agent to fill the tool.

Although the best results may be expected on specialised structural foam machines, it is possible to process normal grades on conventional moulding machines, provided they are fitted with mechanical shut-off nozzles to prevent premature expansion of the blowing agent. Fast rates of injection need to be used in order to prevent the blowing agent expanding before the cavity has been filled. Premature expansion may also occur if the back pressure is too low. It is becoming increasingly common to incorporate glass fibres in foamed plastics to increase the rigidity still further (see Table 6).[2] Foaming agents are normally incorporated in the base polymer and may

TABLE 6[2]
FLEXURAL MODULUS VALUES OF REINFORCED AND UNREINFORCED THERMOPLASTIC FOAMS, kgf/cm^2 (MPa)

Polymer	*Foam*		*Foam* + 20% *GF*	
	kgf/cm^2	(*MPa*)	*kgf/cm*^{2}	(*MPa*)
Polystyrene	21 000	(2 060)	53 500	(5 246)
ABS	9 000	(883)	53 000	(5 197)
Polyethylene	11 000	(1 079)	28 000	(2 746)
Polypropylene	8 500	(834)	28 500	(2 795)
Polycarbonate	21 500	(2 108)	43 000	(4 217)

be of materials such as azodicarbonamide or by direct injection of nitrogen gas.

Applications include fan shrouds, furniture components, pallets, garden frames and television cabinets. A common feature of unreinforced structural foamed materials is a rather rough surface finish which can be described as a 'swirl effect'. This is not so evident with foamed glass fibre-reinforced nylon and it is possible to produce mouldings with a fairly flat smooth finish.

5 SANDWICH MOULDING

Sandwich moulding provides an economic route to a range of composite plastics products. Typical of these are sandwich structures with, for example, a solid outer skin and a central core of foamed plastics. The composite mouldings are made in one operation in an injection moulding machine and the process is generally applicable to a large range of materials. The core and skin may be made from the same polymer with the core possibly containing a foaming agent, or different polymers which stick together well. The skin material may contain fibrous or particulate reinforcements or pigments, UV stabilisers or antistatic agents, or any combination of these and other specific additives. The polymer for the foam core can be the cheapest one with adequate processing and physical properties. Polymer types which have been used include the following, either singly or in selected pairs: high and low density polyethylene, polypropylene, polystyrene, polymethyl methacrylate, ABS, PVC, EVA, nylon, SAN, and polycarbonate. It should be borne in mind that for the process to work satisfactorily the polymers chosen should be roughly comparable in crystallinity and have melting points within 50 °C of each other.

Since, in a large number of instances, the desired properties of a moulding are almost wholly contained within the skin—for example, appearance, and surface properties such as lubricity or even strength—it is possible, when using the sandwich-moulding technique, to incorporate lower grade, or even otherwise reject material, into the core of the moulding. It is obvious that the ideal core material, from an economic standpoint, would be rework of the same polymer as that used for the skin. There have, however, been cases of mouldings made using scrap material.

Whilst the sandwich moulding process was primarily developed as a high pressure process it can also be operated on a low pressure technique using non-expanding constant volume moulds. A shot of skin polymer is injected into the mould first, followed by a shot of foamable core polymer. The quantities are so regulated that at the end of the injection sequence the mould is not completely filled. The foam core component expands, pushing polymer into the extremities of the mould. This creates a foam-cored solid-skinned sandwich moulding.

Mouldings made by the low pressure method offer several advantages:

(a) A machine with low clamping force, large platen area and twin injection units can be used. This will be cheaper to install and run than a high clamping force machine of the same shot capacity.
(b) Conventional 'horizontal flash' moulds can be used. The mould construction is simpler, and therefore somewhat cheaper than for an expanding mould.
(c) Sandwich mouldings of a regular shape with non-uniform cavities can be more easily produced in this type of mould.

The choice of high or low pressure route for the production of a sandwich moulding will obviously be influenced by many factors, including shape, weight saving required, raw material costs, plus tooling and machine costs.

6 CONCLUSION

With the properties of polymers now being more predictable, as the chemist's knowledge becomes more complete, we are probably entering a phase where the chemist's contribution to polymer processing and technology loses its dominance and becomes secondary to that of the engineer. Advances in polymer processing and use will stem largely from changes in the materials, by various means, to suit the engineering requirements of the final articles by modification of the physical properties rather than from the chemical construction of the raw materials.

REFERENCES

1. BRYDSON, J. A. (1975). *Plastics Materials*. 3rd edition, Newnes-Butterworth, London.
2. MANDY, F. (1976). *Plastics and Rubber*, **1** (June), 119–25.

Chapter 6

ECONOMIC MATERIAL USE IN INJECTION MOULDING

G. CHEATER

Head of Productivity Management Group,
Rubber and Plastics Research Association, Shrewsbury, UK

SUMMARY

The usage of polymeric materials has long ceased to be a novelty and they must now stand or fall on their own as materials chosen for a particular purpose. The importance of appreciating the links that exist between choice, manufacture and use as they apply to polymers in particular cannot be over-emphasised and will be discussed in this chapter. Those factors which determine the continued profitability of a product, insofar as it is affected by the material from which it is made, will be discussed. The reasons and the course of reasoning that determine the choice are the target.

1 INTRODUCTION

When once the novelty of a new material or product has passed, its continued use is dependent upon its performance and cost. Performance is dependent upon design and quality and hence on materials; cost is dependent on processing methods and hence, also, on materials.

The choice of material and its economic use can therefore be seen as the paramount decision in the creation of a product. While it is true to say that poor design, marketing or processing may ruin an article's chances, even when materials have been wisely chosen, it is much more difficult for design or processing to overcome or hide the effects of poor or wrongly chosen materials.

The use of polymeric materials has long ceased to be a novelty. They must now stand or fall on their own as materials chosen for a particular purpose. The importance of appreciating the links which exist between choice, manufacture and use, as they apply to polymers in particular, cannot be over-emphasised.

These factors are important when new products are envisaged and they continue to be important when once the product is established and competition arises. With polymers the process is largely determined by the shape of the article; hence, the processing costs are largely fixed and processing economics are mainly found by means of changes in materials, by revising either quantity or grade. Successful defence against competition is therefore dependent upon making the best economic choice in the first instance.

In this chapter it is intended that the reader be made aware of the factors which determine the continued profitability of a product in so far as it is affected by the material from which it is constructed. The reasons and the course of reasoning that determines the choice are the target.

2 CHOICE OF MATERIALS

2.1 The Design Process

All aspects of design and manufacture are subservient to the characteristics of the material. The first requirement in creating an article must therefore be in some way related to an idea of the materials from which it may eventually be produced. It is, however, important that the designer should not limit his thoughts entirely to a single group of materials until he has a reasonably clear idea of the functional needs of the design.

When proposing to utilise one of the polymeric materials, the choice of process is often determined very early in the design stage by reason of size or quantity. It is therefore soon apparent if the article requires a moulding technique and also if it is economic to use the particular process of injection moulding.

In this type of situation the choice of material is then determined by a comparison of the function of various aspects of the article with the physical characteristics of the available polymer compounds. Broad economic considerations may be applied but decisions made at this stage, before detailed design dimensions have been determined, can only be approximations to be confirmed when the final aesthetic and functional shape has been decided.

2.2 Development from Traditional Materials

Frequently the design problem is concerned with a change from traditional materials—metal, wood, leather, fabric, etc.—to a polymeric one. This transition is fraught with hidden dangers that are usually associated with (i) the different ranges of physical characteristics of the material; (ii) the opportunities and/or the necessity of using thinner sections; and (iii) the strength/creep properties of polymers. Metals and timbers have elastic characteristics which give a regular extension under load until a point is reached at which it is apparent that the article is giving way and is about to break. Thermoplastics do not exhibit this behaviour. They will give way progressively, often under very light loads without breaking, and yet they have little ability to recover their original shape when the load is removed.

For these reasons the conversion from traditional materials requires particularly careful treatment if a true assessment of economy is to be made.

Conversely some metal casting processes do not permit articles to be made with thin sections and as a result articles are made thicker than their service duties require. Simply to copy the original shape for the plastic version involves both a waste of materials and a waste of machine time. A word of warning is, however, necessary. Articles are often used in a way that is additional to their designed function. For example, it is said that when the front wings of London's buses were changed from metal to rubber it was overlooked that when changing the route indicator it was usual for the conductor to stand partly on the wing. The rubber wing did not have the necessary rigidity which existed in the steel version as a matter of course. For this reason, if for no other, it is important to adopt a systematic approach to design *before* choosing the material. Furthermore, because material quantity and process cycle time are related, the attainment of optimum economic production is usually achieved when minimum cycle times are obtained as a result of using the minimum weight of the injected material (shot weight).

One approach to this task is to reduce the product to its functional minimum, i.e. the product is treated as a mixture of holes, surfaces and links, these links joint the surfaces and the holes together.

If one considers the simple bracket shown in Fig. 1 its design is obvious; if it is made from steel its strength will probably be appropriate for its size. Convert it to a plastic, say, polyethylene, and the design is immediately ineffective. One approach to the problem is to consider the bracket simply

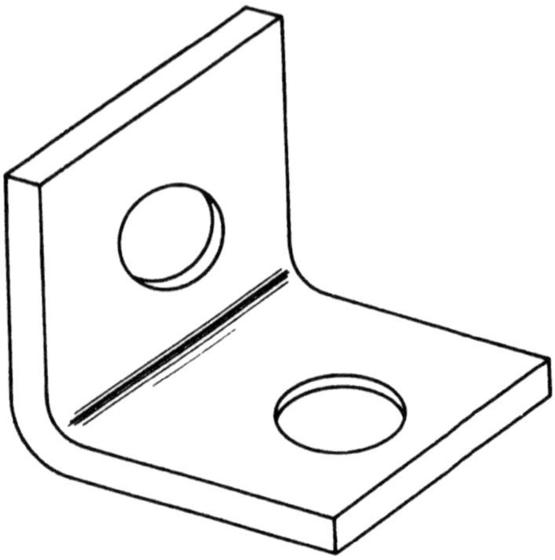

FIG. 1. Metal bracket.

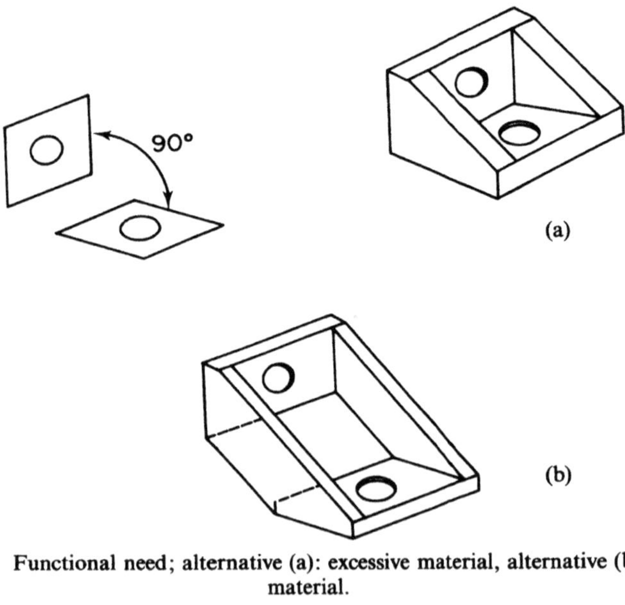

FIG. 2. Functional need; alternative (a): excessive material, alternative (b): less material.

as a means of supporting two holes and keeping them in a rigid relationship at an angle of 90° (Fig. 2). Various alternative configurations then become apparent because one is considering operating functions and not a product.

When once this approach has been applied, and a minimum material content has been determined, it is then the time to consider aesthetics and the manufacturing process. Often it will happen that the functional design is both aesthetically satisfying and suitable for injection moulding, simply because a minimum of material is being used.

Further economies to alternative (b) may exist, if methods of lightening the side webs can be adopted, or if, after due consideration of the whole assembly, the bracket is found to be unnecessary because the joint has been eliminated.

Having determined the minimum functional shape one is then free to consult standard physical data, to choose the material and to calculate the final dimensions that will give satisfactory strength.

3 THE INJECTION MOULDING PROCESS

Thus far, the consideration of design has been applicable to each and every manufacturing technique and material. The special attributes of the injection moulding process, which prompt the need for economic use of materials, are made apparent when one considers the flow of material, starting with the purchased polymer (Fig. 3).

Polymer is usually purchased 'ready for use', although many companies find further economies by undertaking their own compounding or colouring. In the moulding process sprues and runners are formed in order to convey the material into the product cavities in the mould. In general terms the larger the number of cavities the larger will the amount of material required for the mould runners become. Likewise, as the product becomes smaller, so the ratio of material in the runners to that in the product becomes larger. At first sight, because thermoplastics may be regranulated, the economy of material may not seem to be affected, but machine plasticising capacity has been consumed and labour and power will need to be used to recycle the unwanted runners.

It is necessary to consider the process separately for thermoplastics, thermosetting materials and rubbers, in order to understand the material economy of the process.

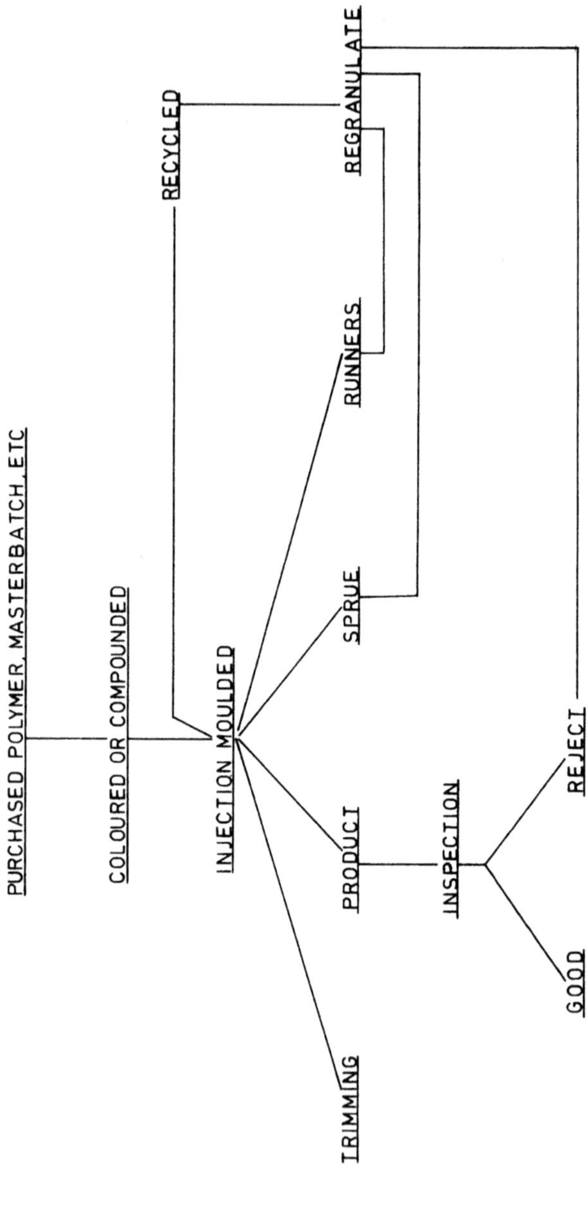

FIG. 3. Injection moulding material flow.

3.1 Moulding of Thermoplastics

Moulding a thermoplastic article involves heating to a 'melt', injecting it into a mould, cooling it in the mould and removing it from the mould. The time taken to prepare the melt is a function of the material: its ability to be rapidly heated, the quantity of heat that is available, and the quantity of the material needed for each injection. Optimum production is obviously achieved when the quantity of material is reduced to a minimum. This quantity comprises not only that which is in the finished product but also all the material that has to be converted to a melt, i.e. that needed for the finished product plus sprues and runners. The smaller the shot weight the shorter will be the time needed to produce a melt from the granules.

Speed of manufacture is also affected by the ability of a polymer to take in heat and by the amount of heat needed to raise the polymer to its melt

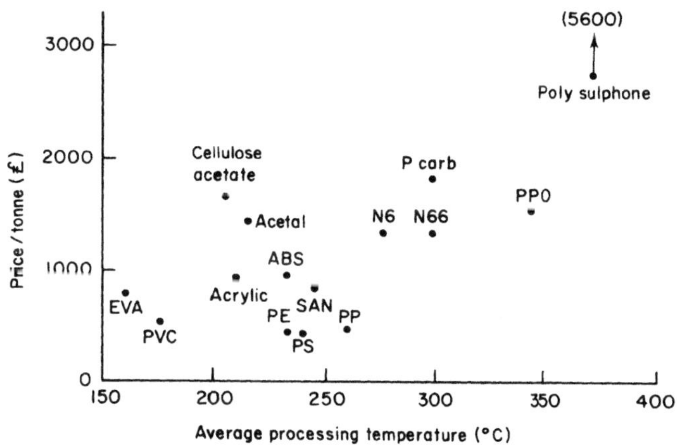

FIG. 4. Relationship between price and processing temperature.

temperature. Higher melt temperatures are usually associated with polymers which have better physical properties. These are also the polymers that are more expensive, and from both cost and physical performance considerations encourage a more careful design procedure. The temperatures shown in Fig. 4 compared with the relative polymer costs illustrate this relationship more clearly.

The in-built ability of thermoplastics to be recycled to varying degrees encourages a lax approach to material usage. It is said that excess runners, sprues, and faulty mouldings can be recycled so material is not wasted.

Whilst this is true of some polymers and some products the number of times that some materials can be recycled is severely limited due either to loss of colour and appearance or of physical properties. Again it is the more critical applications, usually using the expensive polymers, which are more affected by recycling, and the material must soon be downgraded to less severe applications.

It can be shown that by increasing the section of a product, and thereby decreasing the physical demands on the polymer, a higher amount of regrinding may be permitted and the overall economy of the operation thus improved. Such decisions are complex but the fact that they are considered valid indicates the degree of importance which is attached to material economy among manufacturers and users of thermoplastics.

3.2 Moulding of Thermoset Resins

Traditionally it is accepted that recycling of thermoset resins is not done. The fact that they are cured in the mould makes the regranulated material less valuable and less amenable than virgin polymer. However, when it is realised that most thermoset mouldings are a blend of resin and filler, the sense in re-introducing comminuted scrap becomes apparent. Increasingly, moulders of thermosets are examining the opportunities of material economy through regranulation, and are using a percentage of the regrind in place of the filler content. By adopting a policy of recycling and the attitude that scrap is of value there is engendered the idea that there is a virtue in reducing the initial use of the material. The criteria of thin sections is not as critical in thermoset mouldings as it is in thermoplastics; 'sinks' are not as obvious. Nevertheless the same principles do apply and cure times can be reduced by careful design, which reduces section and, thereby, material consumption.

Many articles have been written about the optimum moulding blank size in compression moulding but the tendency still exists to add a little extra to make sure that the mould is filled. One is inclined to say that this happens only when powders are fed directly into the mould and then the operator is blamed. Similar situations arise in better organised factories. The man who calculates the moulding weight wants to make sure! One may imagine that by having the process automated, as is the case within an injection moulding machine, this source of waste is avoided. The reverse is the case, since overpacking is possible simply at the turn of a knob. Mould shot weight control is probably the most effective way of ensuring a good moulding of consistent quality while at the same time keeping a watch on material consumption.

3.3 Moulding of Rubbers

The development of synthetic rubber blends and more recently the introduction of thermoplastic rubbers make the writing of general but meaningful comments difficult. The injection moulding process is similar if normal rubbers are considered as a special type of thermoset and thermoplastic rubbers as a type of thermoplastic. Here the similarity must end and economy must be considered at a much earlier stage for normal or traditional rubbers. The interaction between compound formulation and cure time allows the injection moulder of rubbers freedom either to spend more on the ingredients and reduce time in the machine or vice versa. The opportunity either to shorten or to lengthen machine time is an important realisation, since the economies are by no means always effected in the same direction. Cure ingredients are generally the most expensive part of a compound and the overall compound cost can change quite markedly while very little machine time economies are achieved.

The necessity for reducing waste is also more important with rubbers, since their elastic properties generally render comminution more expensive and avoidance of cross-contamination during re-incorporation into the compound will require close administrative control.

4 MATERIAL RECYCLING

4.1 Economic Considerations

The economic balance of material recycling is constantly changing as the market price for raw material in the various world exchanges fluctuates. Material recycling prices are subject to the demand for materials. Their costs are also subject to the type of accounting procedures employed. Either such materials may have their value ascribed to waste or they may be regarded as virgin materials. They cannot be both. Neither can they be waste one instant and virgin the next! Thus, very clear thinking must apply to the recycling of materials. For these reasons the term 'waste' will be avoided and the phrases virgin or used material will be employed.

Let us consider the case of a hypothetical material which is being processed in such a way that 90% of all the material passing through the process is converted to good saleable products. The cost of manufacture is made up of material plus labour plus machine costs and this will result in N articles plus $N/9$ used material. The remaining re-usable $N/9$ material (i.e. 10% of the input) can (a) be regarded as of no value, its cost divided among the N articles, and material purchase reduced accordingly, or (b) it can be regarded as retaining virgin value and the cost of the N articles reduced accordingly, or (c) it can be ascribed any ratio between these extremes. One reasonable alternative is to ascribe the value that it could command in the open market if sold as 'scrap'.

Whichever accounting procedure is adopted, the effect on the overall situation is similar. True savings can only be made once, although products that are able to be made from a larger proportion of used material should recover more profits for the company.

For detailed descriptions of the methods of recycling plastics and rubbers the reader is referred to such works as reference 1 (Chapter 22) and references 2 and 3. For the purposes of this chapter mention will be made of the two classes of recycling as they affect the material's re-use.

4.2 Regranulation

Comminution to random chips by a variety of 'cutting' and 'hammer' mills is the most usual method of converting waste products (e.g. scrap mouldings, sprues and runners) to a form suitable for re-introduction into the production process. Many moulding departments have appropriately sized machines at the side of the moulding machines and so the material is recycled by the operator and fed straight back into the machine. Where acceptable from the product quality viewpoint this method is probably the least expensive and the most reliable. Where this is not possible, e.g. on fully automated lines or where the used material percentage is too high, regranulating centres are used to provide a source of used material for metering to production.

The production of an irregular, random chip can present a variable within the feed and while it is cheap to prepare, and represents the least amount of heating/cooling of the material, the different feed characteristics of regranulated chips (containing a proportion of 'fines' as compared with those of regular granules of virgin polymer) can cause both feeding problems and variable product quality. For these reasons, where high percentages of sprue and runner occur it is often recommended that they should be subsequently processed into a regular granule before re-use.

4.3 Re-extrusion

Re-extrusion was so named because it entailed an extrusion process after the grinding operation in order to obtain a regular size of granule. Today there are other techniques available to produce either regular cylinders or cubes of recycled material. However, the cost of this extra operation, which involves a heat cycle, must be balanced against the avoidance of material feed difficulties.

5 TOOL DESIGN

A considerable degree of freedom is surrendered when the tool design is determined. For this reason the importance of an organised approach to tool design cannot be overstated. Not only from the material economy point of view is it important; tool design also determines the choice of machine, the amount of labour and the amount of power used. Material economy is largely associated with minimising the amount of material which is to be recycled by careful consideration of the tool design. The methods of minimising the amount of regrind fall into three broad groups: (i) cavity planning; (ii) heated mould runners; and (iii) reduction of spruc size.

5.1 Cavity Planning

A complete book could be written on this subject; indeed the writings are already legion (on, for example, ways of ensuring the most compact location of the required number of cavities), and it is not proposed to add to them save to make the comment that this is the first and cheapest method to be considered. It will usually cost no more in tool manufacture and the material eliminated may permit faster production via shorter plasticising requirements and more speedy injection.

5.2 Heated Mould Runners

By inserting localised cartridge heaters near the runners they may be prevented from cooling with the product and so may be ready to form the next shot. In this way the need to regranulate runners is considerably reduced; the limitation is however that of physically locating the heaters near the runners. Inevitably this approach adds to tooling costs and to the incidence of maintenance faults. These costs must be balanced against the advantages that arise through reduced use of regrind.

5.3 Reduction of Sprue Size

Similar savings can be obtained by designing moulds so that the injection nozzle can penetrate into the mould close to the level of the runners. This approach is usually more easy to adopt than the hot runner approach and, depending upon the tool size, will provide worthwhile economies with very little, if any, extra cost and with no added maintenance hazards.

6 MATERIALS HANDLING AND STORAGE

6.1 Materials Handling

Economic handling of materials is a function of the plant arrangement. It is also an important factor in the efficient use of materials.

The injection moulding machine is an automatic machine requiring little human interference. The operator is simply required to keep a hopper filled with polymer and to ensure that the moulding is ejected. It is strange, therefore, how little attention is paid to the method whereby polymer is taken from stock to the machine and transferred to the machine hopper. Spillage in polymer handling, split bags, poor accessibility to the machine hopper: all these factors, and others, have been shown to be responsible for measurable quantities of waste materials.

Polymer drying is an extension of this handling problem. The use of inefficient and ineffective drying equipment, which allows material to be used even though it is not in the appropriate condition, is indefensible.

Polymer suppliers give concise instructions for using their materials and these usually represent the minimum safe requirements. By avoiding or ignoring them the product is likely to have inferior properties. If these reduced qualities are adequate the question must then be put: is it possible to use a less expensive material or one having less critical processing conditions and using that material correctly?

6.2 Storage

Material economies commence with the purchase and delivery of a consignment. By converting liquid assets into material stock, the base line, from which production economy emanates, is drawn.

The method of delivery, for example, whether by bulk road tanker or in waterproof, 25-kg bags, determines the purchase price, the total investment and the input handling equipment that is needed within the factory.

Ultimate flexibility is obtained by using palletised material packed in 25-kg sacks but these are vulnerable to damage and require individual handling at the machine. It is therefore reasonable to consider whether by storing in larger containers some parts of production may be more easily managed.

Costs of keeping polymer stocks at a correct ambient temperature are becoming of increasing concern as fuel prices rise and as more use is made of 'engineering' polymers (these require close control of moisture content). Warehouses should therefore be equipped with controls that maintain temperatures above the dew-point, thus preventing condensation.

6.3 At the Machine

Clearly the most economic use of material at the machine occurs when spillage is eliminated. Automatic delivery systems from bulk containers in the polymer stores direct to the moulding machine are available and successful systems exist in which a basic range of three or four polymers is available at each machine. Problems can arise through leaks in the ducting and indeed do arise with the need to change polymers at the machine. These latter difficulties are largely due to lack of forethought. The number of portable vacuum cleaners that are to be seen in moulding departments is usually abysmally small and it is not surprising that when hoppers are cleaned material is wasted.

Where 25-kg sacks are delivered directly to the machine the arrangements are much worse. Advantage is taken of the operator's agility so as to avoid the necessity of providing an easy and planned method of opening a sack and feeding its entire contents into the machine hopper. Rarely are machines delivered with any provision for feeding and usually no further provision is made by the purchaser. An arrangement of steps is simple to provide, as are the various types of hopper loader/elevator units, and these often incorporate polymer drying units.

6.4 The Finished Product

For an obscure reason, material value is not conveyed to those who handle or process the material. The value of a finished article is usually an even more closely guarded secret. Publication of these values to those who handle the materials or products cannot but improve the situation and induce a greater feeling of caring.

When removed from the machine the material has been used and further savings cannot be made, but there is still plenty of opportunity to create waste. The moulding is still soft, it will mark and distort easily and become useless. Provision of correct handling procedures and suitable containers is probably one of the more important ways of economising in material use.

7 COMPETITION

Discussion of any topic concerning economic value must be incomplete without mention of the effect of competition in the market place. In this respect it is impossible to determine whether the optimum material economy has been achieved.

The many ways in which material may be consumed, whether planned or unplanned, make material economy the most fruitful hunting ground for savings. This includes the decision of whether or not to change the material grade.

REFERENCES

1. Beadle, J. D. (Ed.) (1971). *Plastics Forming*, Macmillan, London.
2. Blow, C. M. (Ed.) (1971). *Rubber Technology and Manufacture*, Butterworth, London.
3. Penn, W. S. (Ed.) (1969). *Injection Moulding of Elastomers*, Maclaren, London.

Chapter 7

TRAINING AND EDUCATION

C. D. BARNETT

Senior Training Adviser, Rubber and Plastics Processing Industry Training Board, Brentford, Middlesex, UK

SUMMARY

It is very costly not to train and equally costly to do so for the wrong reasons. A company cannot completely sub-contract its training to outside bodies because they cannot know the company's practices. In general the industry does not plan sufficiently in advance to meet its manpower needs, nor does it train very well.

The facilities for the training of craftsmen, technicians and graduates are reviewed along with the availability of short courses for the training of condition setters and other workers.

There is a need to create a career structure for the industry if it is to attract the desired level of abilities to the injection moulding sector. To hold them will require enhanced status and the type of career development that a better educated workforce will expect.

1 INTRODUCTION

Training and education are both costly and time-consuming but it is even more costly to disregard them. On the whole, training in the injection moulding industry is not carried out very well.

When a training officer is appointed, the company often considers that he is responsible for all the training that then takes place. This is not, however, the case. Departmental managers are responsible for the output of their departments, including the development and utilisation of the resources to maximise that output; these should include the human resources and the manager is therefore responsible for the training of his work-people. The training officer is a facilitator and will assist the manager in writing job analyses and drawing up training programmes, but they remain the responsibility of the manager.

A company cannot completely subcontract training to an outside body; for example, a setter can be sent to the Rubber and Plastics Processing Industry Training Board's Injection Moulding Training Centre at Cardiff to learn condition setting. While there he will learn about materials and about how these are affected by moulding conditions. He or she will be taught from first principles and will learn concepts which will be tested out on the plant in the Centre. However, when the setter returns to the company he or she should still undergo a systematic programme of training for familiarisation with the particular machines, moulds and grades of materials used within the company.

Training methods and further education courses have changed considerably over the last 15 years and the courses offered by polytechnics and technical colleges will change at an increasing rate over the next 5 years.

The technology and the structure of the injection moulding industry is also developing and changing at a fairly rapid pace. More and more companies are going over to systems requiring less people but more technical expertise to cope with the introduction of sophisticated machinery and moulds. Many engineering companies are setting up in-house moulding facilities instead of relying on the custom moulder.

There has long been a shortage of skilled labour available to the injection moulding sector of industry, mainly because of the lack of long-term planning of human resources in the past. A realisation of this dawns when a company advertises for, and cannot get, staff such as moulding technicians, mould makers, mould designers and estimators. When this occurs it is about 5 or 10 years too late because it takes this amount of time to train such staff.

One of the major assets in the injection moulding industry is the human asset. This may well prove to be as expensive as the machines and moulds used by the company and certainly has to figure in any investment decision made by that company. It is no use deciding to buy new machines, ancillary equipment and moulds if, at the same time, consideration is not given to the training of people capable of setting and maintaining the equipment in good condition and of getting the maximum utilisation from the investment.

The training and education of the human resources of the company are equal in importance to the marketing function and the terotechnology concerning the machinery. Indeed, none of these areas is given sufficient emphasis in the injection moulding sector.

When a company decides to spend, say, £100 000 on new plant and moulds, it is surprising when the same company turns down an investment of 2% (£2000) in the training of staff to operate the equipment. In early studies carried out by the Engineering Industry Training Board (EITB) it was found that in the better organised and aware companies training and education costs comprised, on average, 4% of their operating costs. How many companies consider this element when budgeting and how many have even considered Human Asset Accounting?[1] In the past, many companies have developed their injection moulding staff by exposure methods, which are very lengthy processes and are based only on their own practices, many of which are misfounded. It is often said, 'We don't need a materials technologist because, if we are in trouble, we get in touch with the materials manufacturer who will bale us out'. Companies which hold this view suffer during materials shortages because no one in the company has sufficient knowledge to select alternative materials. This is even more apparent in a period when materials manufacturers are cutting back on technical services. Many more say that they cannot get skilled maintenance fitters. All of this indicates a need for better planning over the longer term and calls for a greater investment in the development of the human resources of the company.

2 THE NEEDS OF INDUSTRY

Several organisations, including the British Plastics Federation (BPF), the Chemical National Economic Development Organisation (CNEDO) and the Rubber and Plastics Processing Industry Training Board (R & PPITB), which take both a macro- and a micro-view of the industry, have identified shortages in the supply of mould makers, designers, estimators, etc. Companies are also continually looking for injection moulding setters and moulding technicians. I will look at the training of these as separate items later in the chapter and will examine, initially, the need for a career structure in the industry and the inter-relationships of the various posts.

3 TRAINING FOR SPECIFIC JOBS

3.1 Craftsmen

Schemes for the training of craftsmen were developed in the late 1960s by the EITB which introduced the concept of 1-year training off-the-job in a training centre, where the trainees receive a grounding in all the engineering-based crafts such as machining, fitting, sheet-metal work, welding, pipework, electrical wiring, and so on, and progress later in the year towards an eventual specialism in mechanical or electrical engineering, fabrication, vehicle building, welding, etc. This system is now a well-established practice, accepted by most companies training engineering craftsmen. At the end of the year the trainee gains a certificate in basic craft skills. During the year the trainee attends the training centre for 4 days per week for 48 weeks; the other working day is spent at college on a course of further education.

The off-the-job centres are available in most areas of England, Scotland and Wales, and may be housed within a company, in a college, on the premises of a Group Training Association or in centres set up and run by the EITB. In 1974/75 there were about 19 700 1st-year, off-the-job trainees under training, according to statistics published by the EITB.[2] In many areas there are few unfilled places in the centres so that companies now need to recruit early and to reserve places for their trainees by about May of the year in which it is intended to commence the training; in most centres, training starts at the end of August or the beginning of September. Following the 1st-year training the trainees return to their respective companies and are trained by a modular system whereby the company selects the most appropriate modules from the list available.

Figure 1 shows part of the EITB module system.[3] For instance, for a mould maker the appropriate modules would possibly be H.1: Machining for Toolmaking and Experimental Work, followed by H.22: Diemaking. Maintenance fitters and electricians also have appropriate modules to follow, such as J.1 and J.21 plus J.2 and J.22. Each module takes about 1 year to complete and has a skill and training specification which sets out the skill areas to be covered and also supplies a record of what has been experienced. During the training, the trainee keeps a log-book in which he records the various jobs and experiences encountered during his work. All the training is given while the trainee is carrying out actual company work, which must be carefully selected to ensure complete coverage of the training specification. During the module, about eight jobs (one every 4 to 6 weeks) will be assessed by the company's staff as phase tests. These will be selected to cover broad areas of the training specification and the results, along with the trainees' sketches and write-ups, are recorded in the log-book. The trainees will continue with 1 day per week at college studying an appropriate course during modular training. At the end of two successfully completed modules, which are inspected by a Training Adviser from the appropriate Industry Training Board, a Certificate of Craftsmanship will be issued to the trainee.

When recruiting certificated craftsmen, companies should ask to see the prospective employee's craft certificate and the craftsman log-book. The certificate will show which modules have been completed and the log-book will indicate the type of work undertaken during training, thus giving the company proof of a minimum level of skill and also some knowledge of experience gained. If the trainee has undergone more than two modules he will have received an endorsement certificate for each additional module.

Each module has a printed manual which is available from the EITB Publications Department, 54, Clarendon Road, Watford WD1 1LB, Herts. The manuals are very well produced and are likely to be of use for retraining mature workers in another area of skill, for instance in retraining a press tool maker as a mould maker.

3.2 Technicians

The term in itself has little meaning and many arguments ensue from attempts at definition. However, if we say that a technician will do the type of job which requires some scientific and mathematical training and will need the ability to communicate ideas, not very many will argue with such a statement. To carry this line of thought forward, if we then say that the technician's job will include communication, diagnosis, analysis, planning, controlling, designing, and implementing change under the direction of a scientist or technologist (someone with degree-level knowledge and ability), this should set the scene on the technician. Not all technicians will perform all the above tasks, but all their jobs will include several of these elements.[4]

Another complication is that not all the jobs falling into the technician category are first-appointment posts. This means that a job like that of estimator is hardly likely to be the first post of responsibility filled by someone at the completion of, say, 4 years' training. An estimator in the injection moulding industry requires something like 10 years' experience before he or she can cope with quotations for custom mouldings. All the technicians in the injection moulding industry need similar backgrounds, albeit to differing depths, and therefore it is possible to give them all similar general training during the first 3 years. The requirements for jobs such as

3 YEARS

STAGE I Basic training 12 months	STAGE II Module training period 6 months approx.	
FIRST YEAR TRAINING		EXPERIENCE

MECHANICAL ENGINEERING PRACTICES H

MACHINING FOR TOOLMAKING AND EXPERIMENTAL WORK	H.1
TURNING I	H.2
MECHANICAL FITTING I	H.3
MILLING I	H.4
GRINDING I	H.5
COIL SPRING MAKING	H.6
STRIP SPRING MAKING AND FORMING	H.7
VEHICLE FITTING I	H.8
AIR FRAMES AND SYSTEMS FITTING	H.9
INSTRUMENT FITTING I	H.10

MAINTENANCE ENGINEERING PRACTICES J

MECHANICAL MAINTENANCE (AND INSTALLATION) I	J.1
ELECTRICAL MAINTENANCE (AND INSTALLATION) I	J.2
MAINTENANCE OF FACTORY SERVICES I	J.3
ELECTRONIC MAINTENANCE I	J.4
LIFT PRACTICE	J.5

OTHER ENGINEERING PRACTICES K

ENGINEERING WOODWORKING I	K.1

OPTICAL GLASS WORKING PRACTICES O

OPTICAL GLASS WORKING I	O.1

MODULES H, J, K & O

MINIMUM

STAGE III
Module training period 6 months approx.

Necessary qualifications for starting Stage III	Module	Code
	MECHANICAL ENGINEERING PRACTICES	**H**
H.1 or H.3*	PRESS TOOL MAKING	H.21
H.1 or H.4*	DIE MAKING	H.22
H.1 or H.2	TURNING II	H.23
H.1, H.3 or H.10	INSTRUMENT FITTING II	H.24
H.1 or H.3	MECHANICAL FITTING II	H.25
H.1, H.2, H.3, H.4 or H.5	INSPECTION AND MEASUREMENT	H.26
H.1, H.2 or H.4	JIG BORING	H.27
H.1, H.2 or H.4	VERTICAL BORING	H.28/1
H.1, H.2 or H.4	HORIZONTAL BORING	H.28/2
H.1 or H.4	MILLING II	H.29
H.1, H.2, H.3, H.4 or H.5	HIGH SPEED CAN MANUFACTURE	H.30/1
H.1, H.2 or H.4	GEAR CUTTING	H.30/2
H.1, H.2 or H.4	AUTOMATIC TURNING MACHINES	H.30/3
H.1 or H.3	COMPONENT MANUFACTURE	H.30/4
H.1 or H.5	GRINDING II	H.31
H.8	LIGHT VEHICLE FITTING	H.32
H.8	HEAVY VEHICLE FITTING	H.33
H.9	AIRCRAFT ERECTING	H.34
H.3	POWER PLANT FITTING AND TESTING	H.35
	MAINTENANCE ENGINEERING PRACTICES	**J**
J.1	MECHANICAL MAINTENANCE (AND INSTALLATION) II	J.21
J.2	ELECTRICAL MAINTENANCE (AND INSTALLATION) II	J.22
J.3	MAINTENANCE OF FACTORY SERVICES II	J.23
J.4	ELECTRONIC MAINTENANCE II	J.24
J.5	LIFT SERVICING AND MAINTENANCE	J.25
J.5	LIFT ERECTION	J.26
J.21, J.22, J.23 or J24	NUMERICALLY CONTROLLED MACHINE TOOLS	J.27
J.2 or J.4	INSTRUMENT MAINTENANCE	J.28
	OTHER ENGINEERING PRACTICES	**K**
K.1	ENGINEERING WOODWORKING II	K.21
K.1 or D.2	PLASTICS FABRICATION	K.22
	OPTICAL GLASS WORKING PRACTICES	**O**
O.1	OPTICAL GLASS WORKING II	O.21

EXPERIENCE

**If H.3 also Section 8 of H.1.*
**If H.4 also Section 8 of H.1.*

FIG. 1. Structure of craft training.

moulding technicians, mould designers, estimators, tool liaison engineers and technical representatives are such that they all need knowledge and experience of mould making, mould design, moulding, materials, costing and budgeting, and machines and maintenance, in order to appreciate machine utilisation and the problems associated with trying to improve it. They will also need to know about the management control systems such as production planning and control, work study and process and quality control. Following general training the trainees can be given specific training to fit them to a particular job and this may well take between 1 and $1\frac{1}{2}$ years in a first-appointment post.[5]

3.3 Moulding Technicians

This job title is not universally applied and therefore it is necessary to define what is meant by the term. The job involves setting the machine conditions to produce good mouldings in the shortest time using the optimum settings of temperature and pressure. The technician will try out new moulds and produce setting cards for others to work to; trouble-shoot faults which are beyond the ability of the setter; and give advice on moulding to other departments and sometimes to customers. He or she will ensure maximum utilisation of the machines and will often act as a consultant to supervisors in the department. In many cases the job may be carried out by the moulding shop foreman or superintendent; if so, he should have received training and further education before being promoted to the job he holds.

Training for these technicians will mainly be in-company to a planned programme such as that discussed in Section 3.2. Further education is more difficult because there are many alternatives, none of which is specifically designed for injection moulding technicians. There are two courses open which are more specific than others but before talking about the course let us examine what education a moulding technician needs.

He will firstly need to be numerate, he must be literate, and he must be a good communicator. He needs a good understanding of materials and how these are affected by moulding conditions, of how the machine works, and of moulds. The two courses now to be discussed contain all these elements to a greater or lesser extent. One is more advanced in the machinery and moulds area whereas the other concentrates more on the materials and processing side of injection moulding.

The first course is the City and Guilds of London Institute Course No. 255, Mechanical Engineering Technicians (MET). This course is being replaced by the Technician Education Council (TEC) Course based on a unit structure and validated by the TEC A.5 Programme Committee for

Mechanical and Production Engineering. I will, however, talk about the City and Guilds course for ease of explanation. The MET course was in three parts with a common core of mathematics, science, workshop technology and engineering drawing. In addition to the common core the students could select from a large number of specialisms; the specialisms pertaining here are, in Part 1 of the course, 'Non-metallic Materials'; in Part 2, 'Mould Making Technology'; and in Part 3, 'Mould Design and Utilisation'. Not many colleges offered these specialisms but those which did included North East London Polytechnic; Slough Technical College; Matthew Bolton College, Birmingham; Airedale and Wharfedale College, Leeds; and High Wycombe College. With the new TEC courses it is necessary to enquire which units a particular college is offering.

The other course is the TEC 'Polymer Processing Technicians' course. It is also unit-based and allows some specialisms at the Higher Technician Certificate level; these will include units on moulding. The course has a common core of mathematics, polymer science, physical science, engineering principles and an introduction to mould design. The colleges offering this course are also fairly widespread and few in number. Specifically they are National College of Rubber Technology; Polytechnic of the South Bank; Bromley College of Technology; Southall College of Technology; Trowbridge College; Burton-on-Trent Technical College; Ware College; Matthew Bolton College, Birmingham; Manchester Polytechnic; Newcastle Polytechnic; and Napier College, Edinburgh.

Several of these colleges offer full-time courses and all offer the course on a one-day-per-week basis. The Technician Certificate Course takes 3 years and the Higher Technician Certificate a further 2 years on the basis of 1 day per week. Specific details of the courses can be obtained directly from the colleges.

3.4 Estimators

The British Plastics Federation's Moulders Group Committee have, for a long time, been calling for a course to train estimators. It is not difficult to train an estimator if the right person is chosen. Before starting the exercise, someone who has experience of toolmaking, moulding machine setting, mould design, and who has a technical qualification (such as an HNC in Engineering, a City and Guilds Full Technological Certificate in Mechanical Engineering Technicians or Plastics Technicians, or their equivalents) should be selected. Having found such a person it should not take more than about 3 months to make him a reasonably good estimator. Each moulding company has plenty of case studies which such a person could work through and for which he could then produce an estimate of costs. He can then compare his estimates with the actual costs incurred previously in producing the order (all good injection moulding companies have a cost recording system that is factual and reliable). This imposes no risk to the company or to the trainee because he will only be sending the quotation to his tutor within the company.

If, of course, such a qualified person cannot be found in your company, you should consider training people of that calibre for the future. If you start from scratch today you should have a good estimator in something like 10 years' time. Most people, however, are not necessarily school-leavers or qualified engineers, the extremities of the spectrum, but are somewhere in between, in which case it is necessary to analyse their backgrounds and abilities and plan to fill in the gaps. But don't expect a qualified person in the short term. Aim to train to a high standard to avoid disappointment, but not so high that you cannot give the resultant technician sufficient authority and responsibility to sustain his interest in the company. He will also need recognition of status to satisfy his ego. It goes without saying that, since such people are in short supply, he will need to be paid a sufficiently high salary, because if you don't pay him someone else will.

3.5 Condition Setters

When the R & PPITB was set up in 1967, it soon recognised that the greatest grouping of companies that paid a levy to the Board was the injection moulders, which comprised about 20% of all the companies registered. By talking with these companies it was ascertained that there was a dearth of injection moulding setters available to the industry. Therefore the Board took steps to set up a training centre to which companies could send personnel to improve their abilities as injection moulding setters. This facility was provided at the Coventry Skills Centre, Torrington Ave., Canley, Coventry, West Midlands, where the Board installed five machines and organised the courses over a period of several years. This course has now been taken over by the Training Services Agency where it will retrain mature workers as injection moulding setters on a 13-week course organised under the Training Opportunities Scheme (TOPS).

The reason that the R & PPITB discontinued the facility at Coventry was that it had set up a purpose-built Training Centre at Clydesmuir Industrial Estate, Tremorfa, Cardiff.[5] The Board currently offers a two-week course in condition setting for thermoplastics injection moulding, and a similar course for injection moulding and quality control supervisors. Other courses are being developed to include a continuation course—at a more advanced level—for condition setters (thermoplastics); a setting course for injection moulding thermosetting materials and a similar course for rubbers; and a hydraulics course for maintenance fitters.

The Centre is designed and set out to make all production managers envious. The services are gantry-mounted to keep the floor area clear and the equipment is up-to-date. Many of the 12 machines installed will be replaced every 3 years and the Centre will attempt to keep ahead of the injection moulding companies with the introduction of advanced equipment (Fig. 2). For instance, the Centre has one machine fitted with a vented barrel and another has automatic feedback and corrective equipment which is designed to give consistent pressure conditions for each shot. Hot runner moulds are employed, some of conventional design and others using electrical probe systems. The machines available range from about 50 tons lock up to 175 tons lock, giving a good spread amongst the 'middle-of-the-road' machines.

However, the actual machines used are not of paramount importance, because the Training Centre attempts to teach logical fault-finding and correction, together with the establishment of optimum conditions based on first principles and a knowledge of materials. The courses assume little knowledge of materials and of how these react to processing conditions because the Board has found from experience that no matter how long a setter has held that position (even as long as 15 years), he still knows very little about plastics materials and the effect of moulding conditions on them. This means that the experienced setter can usually correct faults once they have occurred, but he rarely has sufficient knowledge to prevent faults from occurring. In the main they are professional 'knob-twiddlers'.

About 50% of the 2-week course is given over to the introduction of background knowledge and to discussing the past experience of other course members. The remainder of the course is practical work under the guidance of one of the six instructional staff employed at the Centre. Typical syllabuses are shown in Fig. 3.

The trainees will return to their companies lacking the specific training needed to operate efficiently in that particular company. It is obvious that no off-the-job training facility can possibly cover the equipment and practices particular to that company. It is therefore necessary for a company to draw up its own programme of in-company training to supplement that gained at the Board's Centre.

After a period of 6 months has elapsed following attendance on a course at Cardiff (plus the additional 6 months' experience), a setter may return to the Cardiff Centre to take an examination under the auspices of the City and

FIG. 2. The Rubber and Plastics Processing Industry Training Board's Training Centre in South Wales has a floor area of approximately 90 m^2 and is equipped with a comprehensive range of injection moulding machines.

Supervisors	Condition setting	Content of courses
	●	Induction
●		Elements of production
	●	General safety, Health & Safety at Work Act
●	●	Machine and mould safety
●	●	Screw thread recognition
●	●	Role of the setter in cost reduction
●		Role of the supervisor in loss control
●	●	Slinging and lifting
●	●	Machine construction
●	●	Clamping methods
●	●	Sequence switches and stops
●	●	Ejection systems (pneumatic, hydraulic and mechanical)
●	●	Basic functions of moulds, 2-plate, 3-plate, hot runner, single- and multi-cavity
●	●	Screws, cylinders and nozzles
●	●	Inspection, care and storage of moulds
●	●	Appreciation of hydraulics and pneumatics
●	●	Fitting and setting ancillary equipment
●	●	Care and maintenance of machines, e.g. greasing and oiling
●	●	The injection moulding cycle
●	●	Material recognition
●	●	Factors affecting shrinkage
●	●	Start-up procedures
●	●	Moulding conditions for HD and LD polythene, polystyrene, polypropylene, nylon 6, 66, 11 and glass-filled, acetal, ABS
●	●	Mould temperature control
●	●	Shut-down procedures and purging
●	●	Pyrometers and thermocouples
●	●	Variables in machines, moulds, and materials affecting quality
●	●	Fault recognition, cause and effect
●	●	Logical trouble shooting
●	●	Condition setting practice, fault finding and correction procedures
●		Quality control and process capability studies

FIG. 3. Typical syllabuses.

Guilds Skills Testing Service. The examination comprises a 1-hour written test (multi-choice questions) and a 6-hour practical test. Success will lead to the award of a certificate for injection moulding setting.

4 FURTHER EDUCATION

4.1 Technician Education Council (TEC) Scheme

Never in the history of education have so many known so little about so much. This could well be an appropriate introduction to courses offered by the Technician Education Council (TEC) or to the council itself. We are in

TEC HIGHER CERTIFICATE					
CU ExM	CU ExM	CU ExM	PW ExM	PW IA	YEAR FIVE
STEC ExM	CU IA	CU IA	CU IA CrU*	CU IA CrU*	YEAR FOUR

	SU/HC† ExM SU/CD	SU/HC† IA	SU/HC† IA				
		TEC CERTIFICATE					
SU/CD† IA	SU/CD† ExM	STEC ExM CrU	STEC ExM	CU ExM	CU ExM	G&C IA	YEAR THREE
		STEC IA CrU	STEC ExM CrU	STEC ExM	CU IA CrU	G&C IA	YEAR TWO
		STEC IA CrU	STEC IA CrU	STEC IA CrU	STEC IA CrU	G&C IA CrU	YEAR ONE

FIG. 4. Example of TEC certificate and higher certificate by day-release study. *Key:* CU, College Units; CrU, Credit Units; ExM, Externally Moderated; G & C, General and Communication; IA, Internally Assessed; PW, Project Work; STEC, TEC Standard Units; SU/CD, Supplementary Units (Certificate/Diploma); SU/HC, Supplementary Units (Certificate/Higher Certificate). *Note:* Units marked with an asterisk might represent such studies as Quality Assurance, Costing, Elements of Supervision, and Work Study. Students on sandwich or block- and day-release courses could receive credit for them if they had appropriate training and experience in industry. Units within broken lines indicate how supplementary units, taken separately, could contribute to another award.

the throes of an educational revolution and most people are unaware of the fact.

The TEC was established in March 1973, following a report of a Committee under the chairmanship of Dr Haslegrave, which studied the education of technicians. The report, published in 1969, recommended the rationalisation of certification for technicians courses and that this should be done by two bodies: the TEC for technical technicians and the Business Education Council (BEC) for commercial and business technicians.

Most people in industry are familiar with qualifications such as the Ordinary and Higher National Certificates and also with City and Guilds Full Technological Certificates. These qualifications will disappear within two or three years, their places being taken by the Technician Certificate (TC) and Higher Technician Certificate (HTC) issued by the TEC.[6]

The new Technician Courses will be of unit structure as shown in Fig. 4, and companies will have to keep in very close contact with colleges to ascertain exactly what course is being offered. This arises because, although TEC Courses will contain a number of standard units, say, 50 % of the total course, colleges will write their own units for the remainder of the Course. This is to allow regional industry requirements to be accommodated.

The certificates issued will present industry with a problem. All the units for which a credit has been achieved will be listed on the certificate by their unit number. There are likely to be many hundreds of units available and no one will be able to know what they all contain. This means that if two candidates apply for a job which is dependent on particular educational standards it will be necessary for some research to be carried out before a comparison of qualifications can be made. Looking on the positive side of the TEC's activities, it does allow special units to be offered to meet particular needs. If the injection moulding industry takes note of this fact and acts upon it we could well have units such as mould technology, mould design, estimating, and injection moulding. To realise this dream, however, sufficient energy must be expended to cause the educationists to sit up and take notice of the needs of the moulding industry.

It is not further education which is at fault when a particular course is not available, it is industry. Take the case of the City and Guilds Mechanical Technicians Course (255). This had numerous Applied Technologies which could be taken as optional subjects in Parts 2 and 3 of the course. In Part 2 Mould Making Technology was available, followed by Mould Design and Utilisation in Part 3. Very few people ever took the course and not more than six colleges throughout the UK offered it. Colleges will not offer courses which industry does not demand. At the same time companies seem to be afraid of going to a college to express their needs for particular courses. Fear is possibly too harsh a term to apply because I do not think there is any emotion involved, just apathy. It is useless complaining to yourself, or to your colleagues about not being able to get this or that course: form a pressure group and demand facilities. But at the same time do not expect to get them on your doorstep because the number of students in any particular geographical location and the cost of capital equipment for specialist courses will limit the provision of such courses.

The injection moulding industry also suffers from a lack of facilities in the teaching field because, in the main, it recruits fairly low academic achievers. With the increase in the applied science, and the sophistication of the whole operation, the industry will need to look to people with higher academic potential for the future growth and prosperity of the sector. There are facilities for the training and education of polymer technicians both on a full-time and on a part-time-day basis, but because of the costly equipment needed to teach polymer processing these are limited to about 12 colleges nationally. It seems to me that, to meet the needs of industry in this training, courses will need to be developed on a full-time or block-release basis and that companies will need to meet the travelling and accommodation costs to allow trainees to reside near the college for the duration of the course.

It would now seem necessary for the interested parties to get together and draw up a comprehensive educational system for the manpower in the polymer industry, these parties being the companies represented by the British Plastics Federation (BPF), the Gauge and Tool Manufacturers' Association (GTMA), the Plastics and Rubber Institute (PRI), R & PPITB, TEC, and City and Guilds. Before this is likely to happen, industry, and the private companies in particular, must stop being apathetic and start using their voices and energy to act as catalyst, accelerator and sustainer to bring about the necessary chain reaction. The introduction of new systems gives the opportunity for all to have a say in the outcome: do not sit back and watch what is happening but get off your seat and participate in bringing about the changes you desire.

The TEC systems of units should meet the needs of technicians in all categories of polymer processing; it will not, however, meet the needs of craftsmen, because these will still be the prerogative of the City and Guilds of London Institute. This is likely to place the mould making company on the horns of a dilemma. In the past, as mentioned previously, Mould Making Technology was an optional subject in Part 2 of the Mechanical Engineering Technicians Course. To ensure that trainees would progress to

Part 2, companies had to recruit people with at least CSE Grade 2 passes in Mathematics, Engineering Science or Physics, and English, or possibly with three GCE 'O' level passes. If such people were recruited for training as mould makers and attended a technicians' course at college, it is unlikely that they would wish to remain 'on-the-bench' as a craftsman for very long because there is not sufficient glamour, pay or prospects to excite the more capable and ambitious.

What can we do, therefore, for the trainee mould maker, upon whom the future of the injection moulding industry depends? Either we recruit people of lower academic potential and suffer the possibility of their not being very numerate, but willing to work as craftsmen, or we recruit brighter people and lose them from the craftsman ranks to become mould designers, production planners, etc. 'Of course', I hear you say, 'we need those people as well'. So we do! But do we necessarily bring them up through the craft route and, if we do, how long will they remain craftsmen before moving to higher things?

4.2 Graduate Training

There are very few professional engineers, materials or processing technologists employed in the injection moulding industry. This seems surprising considering the high cost of the equipment employed and of the materials moulded.

The routes to 'graduateship' are mainly determined by whether study is on a full-time, sandwich or day-release basis. For instance, a Chartered Engineer can only receive such a qualification through a degree in engineering, which is likely to be studied at a university, or at a polytechnic on a sandwich course, by full-time study. This is not necessarily the case with graduates of the Plastics and Rubber Institute because their courses are based on a starting point of Higher National Certificates in Engineering, Chemistry or Physics and these can be studied on a day-release basis.

The in-company training for graduates should be very similar to that of technicians. However, the higher education they receive will be intellectually more demanding, even though it may be very little different in technological content to that of the Higher Technicians' Certificate. Most polytechnics offer HNC courses and many will also run degree courses in engineering and/or chemistry, but few do so for polymer processing. For polymer courses we must fall back to the list of colleges given earlier but must also add some universities to the list, such as Loughborough, Manchester (UMIST), Wales (UWIST), Cranfield, Brunel, Aston, and Imperial College, London.

5 ORGANISATIONS INVOLVED IN TRAINING

5.1 Rubber and Plastics Processing Industry Training Board (R & PPITB)

The Rubber and Plastics Processing Industry Training Board was set up in 1967 (under the 1964 Industrial Training Act), to help ensure better levels of training in the rubber and plastics processing industry.

The Board carries out a Training Advisory role by encouraging companies to adopt a business-related systematic approach to training. Its team of training advisers makes visits to companies to give advice on the identification of company training needs and the preparation and implementation of training plans. These visits are carried out totally free of charge to companies. During discussions the advisers may draw attention to the additional specialist advice the Board provides in such topics as management and supervisory development; craftsman and technician training; industrial relations training; training of young people; safety training; training in small companies; and training in companies with a large proportion of immigrant labour.

The Board has always placed great emphasis on the development of new ideas and approaches which could be helpful to companies. It has, for example, issued a series of guides to the various steps in a systematic training process, for example, the identification of training problems through a survey of training needs; and a training review for the smaller company. It has also published a number of recommendations on training for particular jobs, such as operatives, GRP laminators, craftsmen, managers, supervisors, tyre-fitters, professional engineers, technicians, etc., and some subject guides which have application right across a company's operations, for example industrial relations and safety.

The basic objective of the Board's development work is to ensure that its training advisory service is innovative and imaginative, and that new and useful ideas can be offered even to the best of companies.

The Board serves a wide range of companies whose needs and priorities vary considerably according to the size and nature of their business. The rubber and plastics processing industry comprises some 2200 establishments employing a total of nearly 290 000 people. Approximately 75% of these establishments employ less than 100 people.

There are more than 30 industrial sectors which have establishments within the Board's scope. Among the principal of these are tyre manufacturing and distribution, injection moulding, moulding of general rubber goods, plastics and rubber extrusion, blow moulding, compression moulding, plastics foam manufacture and plastics coating.

5.2 Colleges

Apart from the full-time and part-time certified courses offered by colleges, many also run short courses and seminars of a more specific nature. Most of the colleges previously listed organise short courses, conferences and symposia but mainly on an *ad hoc* basis and they therefore do not publish their programmes very far in advance. Information can be obtained from the individual colleges.

The benefit of short courses is mainly to those previously qualified, who want 'topping-up', or for people such as supervisors who have had little formal education on polymers or injection moulding and who cannot be released for longer-term training. Seminars are usually aimed at the more knowledgeable employees from industry.

5.3 Other Organisations

5.3.1 *Machinery Manufacturers*

Many companies manufacturing injection moulding machines run courses for their customers in the utilisation and maintenance of the specific machines which they manufacture. Such names as Bipel, Netstal, Engel and Demag-Hamilton come readily to mind. Most of the courses will be of about 3 days' duration but details of dates, course content, price and the target people at whom the course is aimed must be ascertained directly from the company or their agents.

5.3.2 *Equipment Manufacturers*

Courses on hydraulics are offered by companies such as Sperry Vickers at the Birmingham training centre and pneumatics courses are available with Maxam Ltd, in Cornwall. Instrumentation courses are organised by companies such as Eurotherm Ltd. Time and space do not allow details of these courses to be covered here but they may be acquired directly from the companies or from the information service of the R & PPITB.

5.3.3 *Injection Moulding Consultants*

Apart from the injection moulding setters course run by the R & PPITB at the Cardiff Centre, and also the short courses run by the machinery manufacturers mentioned previously, other courses are available. Optimum Instruction Ltd has a training centre in New Cross, south-east London which runs modular courses. PERA (Production Engineering Research Association) runs courses in various regional locations, as does a consultancy called Selection and Industrial Training Administration

(SITA). These tend to be of 3 days' duration and are run at an hotel, which means, of course, that the sessions will be knowledge-based.

5.3.4 *The Training Opportunities Scheme* (*TOPS*)
The Training Opportunities Scheme, under the aegis of the Manpower Services Commission (MSC), runs several courses which are of interest to injection moulders.

There is an injection moulding setters' course, of 13 weeks' duration, at the Skills Centre in Coventry. The course is designed for the retraining of unemployed people and it comprises 5 weeks of basic engineering craft skills and 8 weeks of injection moulding setting. The Centre has some five moulding machines ranging from about 1 oz (25 g) to 6 oz (150 g) capacity.

The TSA (Training Services Agency) offers a mobile instructors service and has three instructors who will run in-company courses for injection moulding operators. These courses can be negotiated through local Government skills centres.

The Skills Centre in Handsworth, Birmingham, offers a 26-week course in diesinking. This course is in three parts: basic milling, diesinking, and engraving. For company-sponsored trainees the Skill Centre will allow parts of the course to be taken and this can be negotiated locally.

Other skill centres run courses for specific skills such as fitting, milling, turning, toolmaking, electrical engineering, electronics, etc. These centres are located in most large towns and there are about 100 in the UK. Details of the courses they offer are available at local offices of the Training Services Agency.

6 FUTURE NEEDS

Firstly there is a need to establish career patterns for the industry if it is to attract academically qualified young people on to the industry ladder. With a highly developed technology, firmly based on scientific principles, companies should be looking to recruit young people with good 'O' and 'A' level qualifications. However, these people will not offer their services to an industry that cannot demonstrate that there are sufficient avenues along which to progress as well as goals to achieve, nor will they be interested unless their potential can be realised through good training schemes and suitable courses of further education. A possible structure is shown in Fig. 5.

The industry needs to be more systematic in its planning for the future.

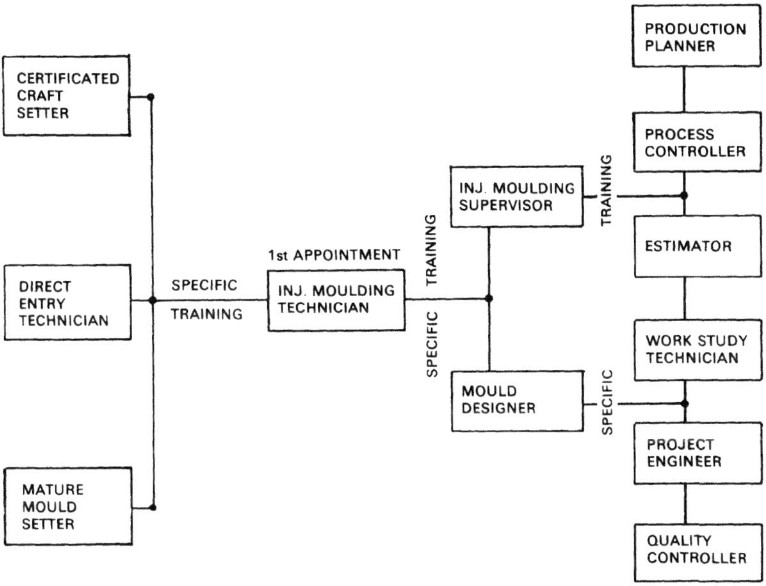

FIG. 5. Career pattern in injection moulding.

This must include the acquisition of equipment, the development of management controls, particularly costing systems, and the development of personnel.

The costs of training are high but not as high as the cost of not training. At the time of writing, the direct cost of external courses will be between £15 and £35 per day in course fees, employment costs during the course will be about £20 per day, and accommodation £15 per day. Thus, external training can cost a company as much as £300 per week. If a company is to be deprived of the services of a key person and pay about £300 per week it is necessary to plan very carefully to gain the maximum benefit from the exercise. As was said previously, companies cannot entirely sub-contract their training so a planned in-company course must be implemented when these trainees return.

7 CONCLUSION

Education and training is a complex subject, almost as complex as business itself, and like business it needs careful planning and evaluation. Many facilities are available to companies for the training of their staff. There are perhaps not as many as companies would desire, and many of them may be located a long way away, but they do exist. One of the major problems arises because companies are not aware of their existence and do not know where to go for information. Advice is available from the R & PPITB Information Service, from the Education Officer of the Plastics and Rubber Institute, 11 Hobart Place, London, SW1W 0HL, or from your local college.

REFERENCES

1. Giles, W. T. and Robinson, D. F. (1972). *Human Asset Accounting*, Institute of Personnel Management, London.
2. EITB (1977). Research, Planning and Statistical Research Paper No. 1. EITB Watford.
3. EITB (1974). *The Module System*. EITB Watford.
4. R & PPITB (1971). *Training of Technicians*. R & PPITB London.
5. R & PPITB (1977). *Introducing Britain's First Injection Moulding Training Centre*. R & PPITB, London.
6. TEC (1974). *Policy Statement*. TEC, London.

Chapter 8

TEMPERATURE CONTROL

P. H. J. Ingham

Marketing Manager, Eurotherm Ltd, Worthing, Sussex, UK

SUMMARY

Commercial plastic materials are organically based and are therefore heat-sensitive. Accurate temperature control of melt processes such as injection moulding is therefore necessary if problems caused by thermal degradation are to be avoided.

The injection moulding process is considered from a temperature control viewpoint and some of the control methods or techniques are described, since it should not be forgotten that good temperature control can lead to materials and energy savings.

1 INTRODUCTION

The injection moulding process is concerned with the efficient conversion of plastics raw material into moulded product of acceptable standards. Some of the parameters which determine acceptability are weight, dimensions, colour and strength, all of which can be affected by the conditions under which the material is processed. Having established the tolerances for these parameters so as to determine acceptability, limits can be set for the conditions under which the material is processed. One of the most important parameters contributing to the correct operation of an injection moulding machine is temperature. All plastics materials can be correctly processed only within a certain range of temperatures which varies from material to material. For some materials and mould types the band is very small and for others it can be quite wide.

Any attempt to define the limits within which the product is acceptable determines the need for some form of control. There are a number of types of control which, if applied correctly, can lead to adequate performance. Significant material and energy savings can be achieved by correctly applying the right type of control equipment. The reliability of the system and the degree of operator supervision required also depend very largely on the balance struck between initial cost, running cost and performance.

It is the purpose of this chapter to examine the injection moulding machine from a temperature control viewpoint and to outline some of the control methods which can be used, together with their advantages and disadvantages.

2 THE PROCESS

2.1 Machine Zoning

From a control viewpoint, an injection moulding machine consists of a number of zones (each equipped with a means of measuring the temperature) and a controller, which compares the measured value of the set-point and controls the heat input to the zone in such a way as to remove any difference between the two. By dividing the machine into a number of zones the different temperature requirements of different zones and their different heat input needs can most easily be met (Fig. 1). For this purpose a

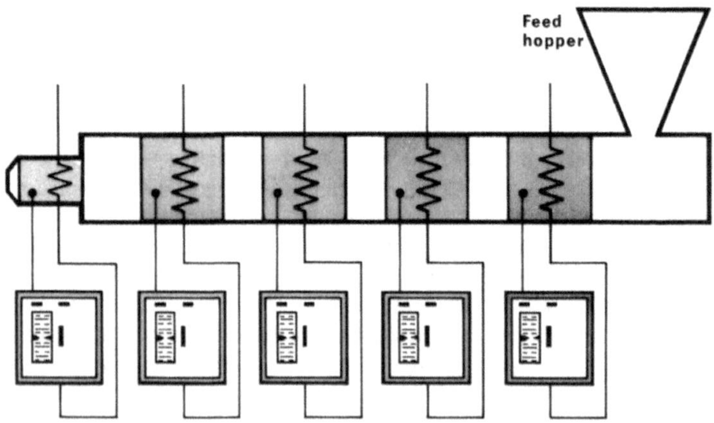

FIG. 1. Typical arrangement of injection machine with four barrel zones and a nozzle heater.

typical small machine may have three or four barrel zones and a nozzle zone. The zones nearest to the material feed hopper are where the plastic is melted and thus require fairly large heat inputs. However, in the zones nearest to the nozzle, the heat produced, by the rise in pressure needed to force the plastic into the mould, means that relatively little additional heat input is required when the machine is running. Indeed, if the machine cycle is very short, with some materials it may be that more heat is generated than is required to maintain the temperature, which will then rise uncontrollably unless some form of additional cooling is applied.

2.2 Thermocouple Location

Considering again the barrel zones: these consist of a metal barrel with wall thickness sufficient to withstand the high pressures produced during the injection cycle. The most common form of heating is electrical and is applied using band heaters strapped around the barrel (Fig. 2). A controller of any kind can only control the temperature at the point of measurement. Ideally this will be as deep into the barrel wall as possible, since it is the temperature of the plastic which is required and not that of the barrel. Plastic is a poor thermal conductor and depending on whether the net heat flow is into or out of the plastic, a thermocouple deep into the barrel wall will register a temperature above or below the actual temperature. If the measuring element is shallow or on the barrel surface, the difference between the measured and actual melt temperatures can be very large. For

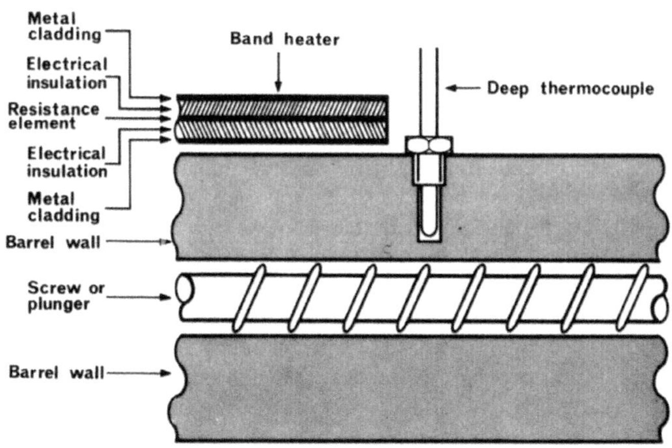

FIG. 2. Section through the barrel wall and heater.

any given conditions of operation there will be a more or less fixed difference between the melt and measured temperatures and acceptable product may be produced. If, however, the conditions, e.g. machine speed or ambient temperature, change, this may give rise to a melt temperature which does not result in the production of acceptable product. It is therefore important to place the thermocouple as close to the melt as possible, i.e. deep in the barrel.

2.3 Temperature Overshoot

The resultant system of an electrical band heater strapped around a thick walled barrel with a deep thermocouple is typical of most plastics processing machinery and presents a number of control problems. Not only must stable control be achieved during normal running of the machine but acceptable start-up performance must also be achieved. The machine must be brought to its normal operating temperature as quickly as possible and preferably with no overshoot. (Overshoot is said to occur if the temperature is rising or falling at such a rate as it reaches set-point that it does not stop there but continues past by some amount before returning towards set-point again; see Fig. 4.)

The basic cause of temperature overshoot in the system is multiple heat-transfer lags, i.e. where the heat generated electrically first raises the temperature of the heater thermal mass and is then conducted to a second thermal mass, thus raising its temperature; it is then conducted from the second thermal mass to a third and so on, until the heat reaches the point of measurement which, as stated already, is as near as possible to the point in the process to be controlled.

In the simplest case of multiple heat transfer only two thermal masses would be significantly involved, namely those of the heater and the load. If the thermal mass of each is about the same, this tends to represent about the worst case for overshoots (and hence controllability). Poor heat transfer from heater to load worsens the situation, since the heater temperature (during start-up, for example) can then become very much higher than the load temperature; when the power to the heater is cut off the final temperature reached (ignoring heat losses and assuming equal thermal masses for heater and load) will be the mean of their respective temperatures at the instant when the power is cut off. Thus, the overshoot in load temperature increases as the heat transfer becomes worse.

A particularly bad case of overshoot (and controllability) occurs where heat is transferred through a considerable thickness of heat-conducting material. This is exactly the situation which is presented by an injection machine barrel with deep set thermocouple. This sort of heat transfer represents in effect an infinite order multiple heat transfer: several minutes can elapse between switch-on of power and a significant change in thermocouple temperature. In fact the response has almost the appearance of a delay (i.e. transport lag) although there is really a considerable difference between this heat-transfer lag and a true delay. During the time of the heat-transfer lag, heat is being fed into the barrel, so that even if the source of heat were switched off at the instant the deep thermocouple began to respond, the thermocouple temperature would continue to rise as the heat energy already fed in distributed itself evenly throughout the thickness of the barrel wall.

A large part of the total lag can in practice be caused by the heat-transfer lag which occurs with a resistance heater, from the heater element thermal mass, via electrical insulation, to the outer surface of the barrel. For the lag through the barrel wall (or for any similar form of heat transfer) doubling the heat-transfer distance results in four times the lag. Iron, from which most injection machines are made, is a rather poor material for heat transfer: for example, similar lags are obtained in aluminium and iron when the distance in aluminium is five times greater.

3 METHODS OF CONTROLLING TEMPERATURE

3.1 Measuring the Temperature

The first item in the control system to consider is the measuring element, of which there are two basic electrical types: active and passive.

The *active* type are thermocouples. These are formed by the junction of two dissimilar metals and give an output voltage proportional to the difference in temperature between the thermocouple and the point of measurement (Fig. 3). The fact that the millivolt output of the thermocouple in relation to temperature is non-linear and that it depends on a stable reference temperature for comparison purposes are factors

FIG. 3. Millivolt output measured is proportional to $T_1 - T_2$ where T_1 is thermocouple temperature and T_2 is ambient temperature.

which must be taken into account in the controller. Thermocouples are very robust mechanically. (This is an obvious advantage in the environment of the moulding shop.) They also exhibit good repeatability from example to example of the same type. The two most common types used in plastic processing are both base metal thermocouples and these are nickel chrome/nickel aluminium (Type K) and iron/constantan (Type J).

The *passive* types rely on having a resistance which varies with temperature in a known manner and thus, when fed from a constant current source, will cause the voltage which appears across them to be dependent upon temperature. Such elements do not require a reference temperature to be generated by the controller. The commonest are the platinum resistance thermometer (which occupies a larger volume than a thermocouple and is more fragile) and the thermistor (which operates on the same principle and has the same disadvantages).

The thermocouple is by far the most common measuring element used in practice. The siting of the thermocouple will depend upon the degree of control required, as will the choice of controller.

3.2 ON/OFF Control

The simplest form of controller provides ON/OFF control of load power. The measured temperature is compared with the set-point and if it is too low, power is applied to the load; if it is too high the power is switched off. In practice there will be a small amount of hysteresis in the controller (mainly so that spurious noise signals on the thermocouple and effects due to mains regulation should not result in rapid ON/OFF chattering of the load power

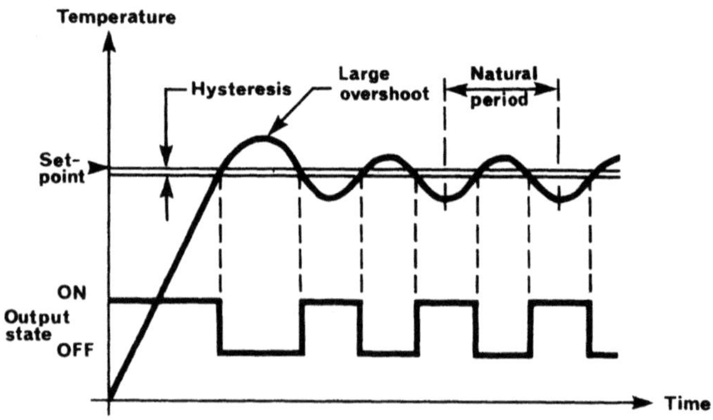

FIG. 4. ON/OFF controller performance showing temperature cycling.

control relay). If the thermocouple and heater are in very close proximity, i.e. there is no appreciable lag, the temperature will cycle with an amplitude somewhat in excess of the controller hysteresis and with the natural period of the system. There will inevitably be some overshoot on start-up because full power will be applied to the load until the set and actual temperatures become equal and any stored energy in the heater will continue to be transferred to the load even after switch-off. It can be seen that if the thermocouple is deep in the barrel (thus measuring the melt temperature more closely) the system lags will be considerably increased and the temperature cycling will be of a longer period and will become much larger. Similar comments apply to the start-up overshoot.

Thus, in the least demanding circumstances, an ON/OFF controller with a shallow thermocouple may give acceptable results. However, with the large heaters required to give short start-up times, the start-up overshoot will probably be unacceptable for all but the least demanding situations and will be worse if account is taken of correct siting of the thermocouple.

The natural period of the system results from a combination of heater power and location, sensor location, and the thermal mass of the system.

3.3 Proportional Control (P only)

If we take an ON/OFF controller and force the switching of the output within the controller itself (with variable mark : space ratio) at a rate which is higher than the natural period, then we have proportional control. As the measured temperature approaches the set temperature, the relay will switch off (for a short time) the power supplied to the load. This point, at which just less than full power is applied to the load, is the lower edge of the 'proportional band'. As the actual temperature approaches the set temperature more closely, less and less power is applied to the load until, when the two become equal, the power input is zero. It is general for the proportional band to be downscale of the set-point, i.e. at set-point the power fed to the load is zero.

The proportional band is usually defined as a percentage of the controller set-point scale span. Since the power applied to the load is proportional to the error or difference between actual and measured temperature (a so-called error-actuated system), it follows that if any power is required to maintain the temperature there must be some error in the system. This error is known as offset or droop (Fig. 5). Since, on start-up, the load power will first be switched off at a temperature below the set-point, the resultant overshoot will be reduced. With a sufficiently large proportional band and sufficiently rapid cycling of the output power (compared to the system's

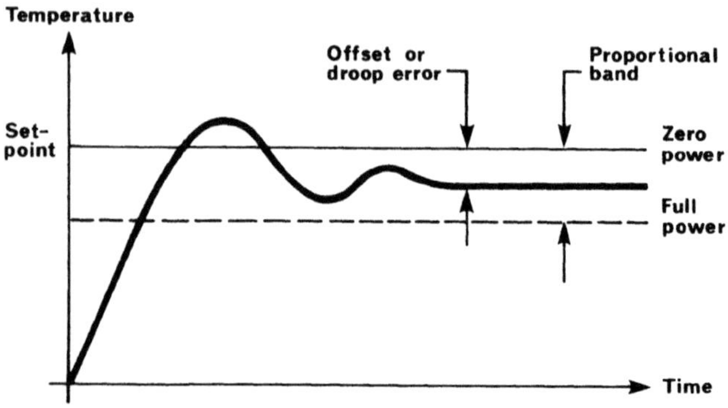

FIG. 5. Proportional control. The proportional band is wide and there is a small overshoot.

natural frequency) the oscillations in temperature will cease eventually. However, this does not necessarily mean that there will be no start-up overshoot in temperature, but only that the subsequent oscillation will decay to zero amplitude.

Further increases in the width of the proportional band beyond that width just resulting in stability (i.e. cessation of a steady-state oscillation) do in fact reduce the initial start-up overshoot magnitude as well as making the unwanted oscillations die away more quickly. Eventually, as the width of the proportional band is further increased, the initial start-up overshoot and subsequent oscillations will probably disappear altogether.

To have a significant effect on overshoot behaviour and subsequent oscillation (sufficient to stabilise the system against continuous oscillation in temperature) the proportional band must have a magnitude or width of the same order as the ON/OFF system's peak-to-peak oscillation magnitude.

Because the ON/OFF cycling time in the proportioning system is made (in the controller design) to be faster than the shortest oscillation period normally encountered, the fluctuation (*not* oscillation) in temperature, which is caused by the proportioning ON/OFF action is very much smaller than the oscillation in an ON/OFF system.

Since the output of the controller is proportional to error, we can assume that, for a particular set of machine conditions, and thus for a particular power input requirement a particular error will result. If the proportional band is wide enough to improve significantly on the start-up performance, the error may be unacceptably large. This can be overcome in part by the addition of an 'offset' control (known as manual reset) with which the proportional band is moved relative to the set-point or zero error. However, any change in machine conditions will result in a different power input requirement and a new control point. Thus, it can be seen that with proportional control the machine temperature will not be held very precisely unless some account is taken of changes in supply voltage (see Section 4).

3.4 Proportional plus Derivative Control (PD)

The initial overshoot in a proportional-only control system could probably be overcome by a sufficient widening of the proportional band but the control would then be too 'sloppy'. A significant reduction in overshoot can be achieved (while still maintaining a proportional band narrow enough to give stable control) by adding to the controller a 'derivative' term (D).

A differentiating circuit is added to the controller and its output, which has a magnitude proportional to the rate of change of the temperature, is added to the signal which is proportional to the error. It is added in such a way as to try to minimise the rate of change. For example, the temperature may be below the set-point and if, it is rising rapidly, derivative action would cause the load power to be reduced to a value less than the proportional output corresponding to that particular error. If the derivative term is chosen to be of the right magnitude, i.e. sufficient to counteract the lags in the system, overshoot can be minimised or even eliminated (Fig. 6).

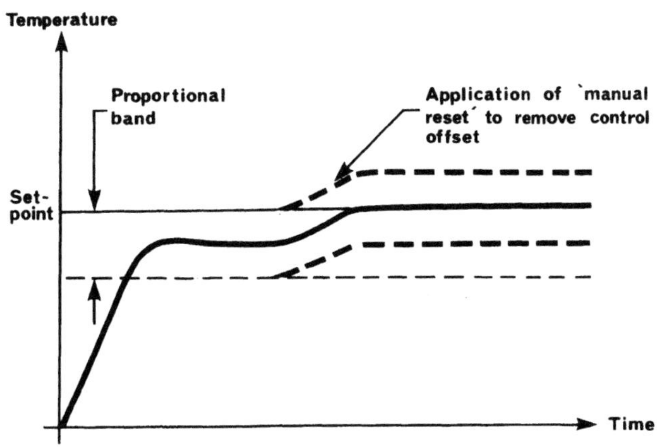

FIG. 6. PD control showing no overshoot and the effect of manual reset.

With an error signal that changes linearly with time, the differentiator output will be of constant magnitude: the time for the error signal to then change through an amount equal to the (constant) output from the differentiator is defined as the derivative time constant.

Derivative action obviously has a beneficial effect during normal machine operation, since the controller reaction to any kind of disturbance (e.g. machine starting or stopping) will be much stronger, resulting in tighter control. Proportional plus derivative control may permit the use of deep thermocouples and a proportional band narrow enough to give tolerable control for undemanding situations.

3.5 Proportional plus Integral plus Derivative Control (Three-term or PID)

The accuracy limitations of a P or PD controller (as described in Sections 3.3 and 3.4) are due to the fact that an error is necessary for power to be applied to the load. The error will vary, depending on the load power requirements, but for steady conditions can be adjusted out by moving the proportional band relative to the set-point (manual reset). This cannot be considered a very satisfactory approach since there are so many external influences on the machine which will cause the error to change.

A means of automatically removing the steady state error is to include in the controller an integrator. This slowly accumulates a small error signal so as to give eventually, from this small error a larger power-control signal which is applied, in such a direction (or sense) as to reduce the error. The action of the integrator output provides a bias signal to the power output stage similar to that provided by the manual reset adjustment mentioned earlier. The proportional error acts in the same sense and this integral of error simply adds to the output power thus applied to the load.

In this way, it is not necessary for a steady-state error to exist (to supply the control signal needed for a given required mean level of output power) because the required power-control signal is now the output signal of the error integrator, which will remain at its fixed level even if the input is zero—i.e. zero error (Fig. 7).

The integral action is 'sluggish', so that some time must elapse before the error 'integrates out'. One way of looking at the integration is that it results in zero bandwidth because the steady-state gain of an integrator is very large, and yet it does not cause the system to oscillate continuously (as would the zero bandwidth ON/OFF controller) because of its sluggish response. In fact, the integral term is a very definite destabilising influence, and must be made sufficiently slow-acting if it is not to be troublesome in this way.

The integral control term is destabilising in nature because it remembers or accumulates the error which has been occurring, and therefore causes the power-output to switch off at a later instant than otherwise: how much later (and therefore how much unstabilising effect this has) depends upon the magnitude of the integral signal—i.e. upon the integral action time or time constant. This is defined as the time, with a given constant error-signal, for

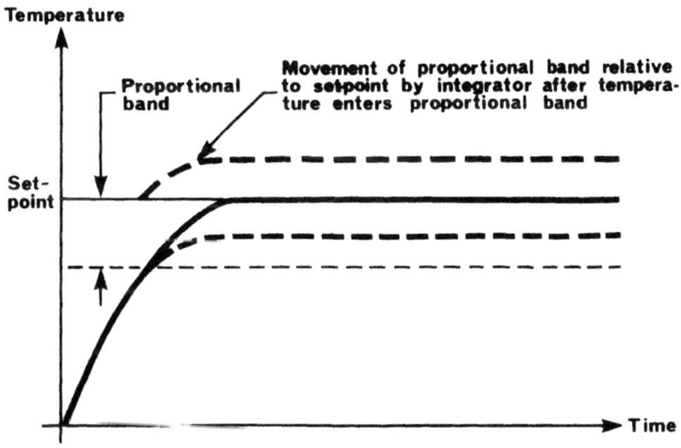

FIG. 7. PID control showing no overshoot and automatic control at set-point.

the integral control signal to become equal in magnitude to the given proportional control signal. Larger integral time constants give less unstabilising effects and the integral time constant must in any case be longer than the natural period (as in ON/OFF control) for severe unstabilising effects not to occur. For this reason PI controllers are not generally applied to plastics processing equipment.

In order to minimise the 'storage' of unwanted error signals (e.g. those occurring when the error is very large shortly after switch-on) it is necessary to include as part of the 'overshoot' inhibition system some means of removing the integrator from the circuit until the error is relatively small. This is usually effected by 'shorting' out the capacitor on which the integral information is stored if an output power greater than 100% is being demanded (i.e. if the measured temperature is below the proportional band).

Some control equipment is described as PDPI rather than PID. However, examination of the transfer functions of these two circuit configurations reveals that there is in practice no significant difference between the two, and indeed the double derivative action of the PDPI can lead to inferior noise performance.

4 AVAILABLE CONTROL EQUIPMENT

4.1 Stability and Repeatability

Controllers used on injection moulding machines can be, in the simplest case, ON/OFF with no indication of temperature, or PD and PID with indication. ON/OFF and PD controllers are generally used in low-cost applications where their performance may be adequate. PID control is required in instances where a high stability of temperature must be maintained under all operating conditions (Fig. 8).

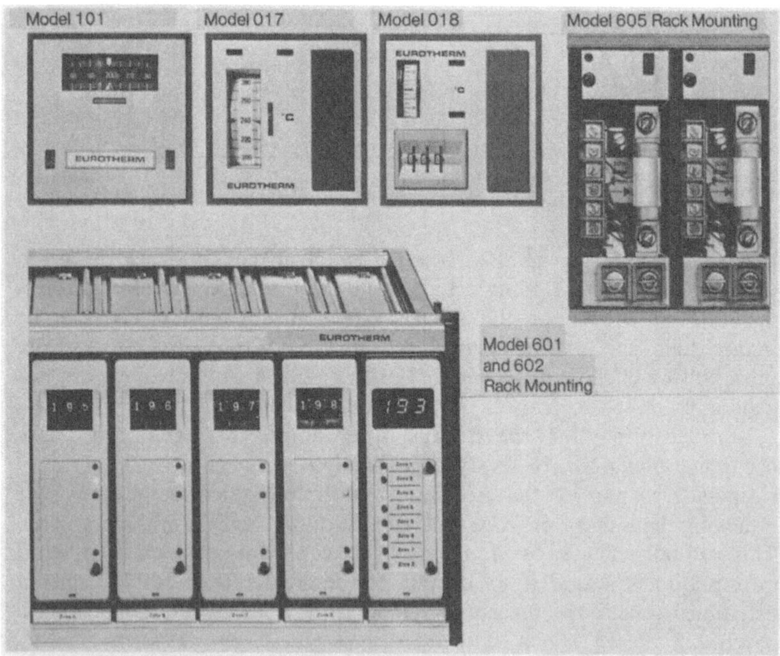

FIG. 8. A range of PD and PID controllers as single units and in a multiple racking configuration with shared digital indicator, together with thyristor units rated up to 50 A (480 V) for mounting inside a cubicle.

Controllers may be fitted with an analogue set-point mechanism (preferably of the drum-scale type, to provide good scale length for accurate temperature setting and reading) or with a thumb-wheel digit switch and error meter. The latter arrangement is particularly useful when set-point repeatability is of importance and also in avoiding interpolation errors due to misreading of scale graduations. Typical set-point accuracies are 0·5 % of set-point span, i.e. plus or minus 2 °C for a typical 0–400 °C controller scale.

This is at least as good as the differences found between different examples of the same type of thermocouple. Set-point accuracy is in any case not as important as stability and repeatability since often the required set-point is arrived at experimentally but must then be maintained accurately.

Cold junction compensation, usually in the form of an electronic reference, is built into the controller and should achieve a 10 to 1 rejection of ambient temperature changes, i.e. the load temperature alters not more than 1 °C for every 10 °C change in ambient temperature. The mains voltage can change quite substantially in an industrial environment and a temperature controller should reject changes of −15 to +10% of the nominal supply voltage.

The controller output stage can be fitted with power feedback. If a measure of supply voltage is used to determine the ratio of load power ON to OFF times for a given control signal within the proportional band, then changes in supply can be used to change this ratio instantly so that disturbances which would otherwise occur are minimised. This feature is particularly important since a 5% change in supply voltage will cause a 10% change in load power and this may give rise to a considerable disturbance in the load temperature, since power is proportional to the square of the applied voltage. Power feedback may reduce this to as little as 1 or 2%, leaving only this small disturbance to be removed by the temperature control loop.

4.2 Controller Output Types

Controller outputs are generally of two types: (i) relay or triac, or (ii) thyristors. Relays and triacs are usually used with external contactors for an economical power switching system. The rate of switching is a limitation and would be chosen to be fairly slow so as to avoid excessive wear on the moving parts. However, if the rate approaches the natural period of the system, significant temperature ripple will result. A compromise has to be found which generally results in typical (ON + OFF) times of 20 sec or so. A contactor has an expected life of around one million operations; thus, it would need to be replaced after about two-and-a-half years if the machine was in use for 8 h each day.

Thyristors, however, because of their 'solid-state' nature, can switch the load power very rapidly without any wear and can therefore offer long life and better control of temperature. They can be operated in one of two ways: phase angle and fast cycle. In the case of phase angle, the thyristors are

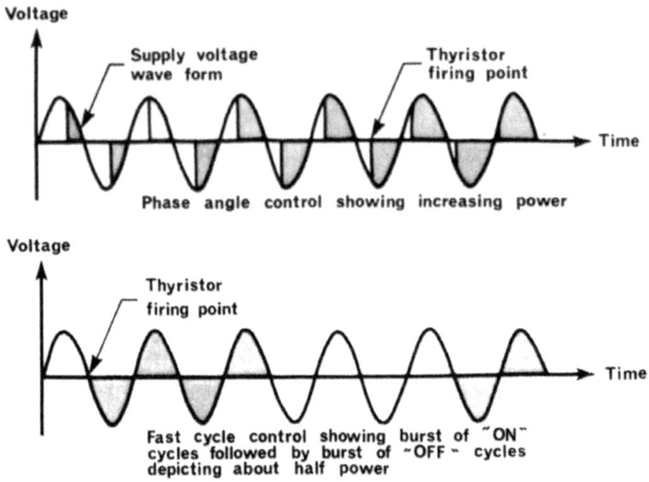

FIG. 9. Phase angle and fast cycle control.

switched on in every half cycle of the mains supply and the load power is controlled by varying the point in the cycle at which switch-on occurs (Fig. 9). More commonly on plastics machinery, fast cycle firing (burst firing) is used. Here the thyristors are switched on at the beginning of a half cycle, when the supply voltage is zero, and remain conducting through several cycles, the power input to the load being regulated by the ratio of ON to OFF time. Fast cycling has the advantage of producing little radio frequency interference or supply waveform distortion. Typical (ON + OFF) times are about 0·2 sec and this results in very smooth control of temperature. Such rapid cycling minimises thermal stress experienced by the heaters, thus prolonging their life. (*Note:* even such rapid cycling would be too slow for a lamp type heater where phase angle firing would be essential.)

5 OPTIMUM CONTROL

5.1 Controller Settings

The combination of controller parameters, i.e. the proportional band employed, the integral and derivative times used, together with the process being controlled, determines whether the process is stable and whether good start-up performance is obtained. In the case of an ON/OFF controller, the overall performance is determined completely by the process, i.e. the power of the heater and the siting of the thermocouple. A shallow thermocouple may give better results than a deep one since this would reduce the heat transfer lag.

In the case of a PD controller, the proportional bandwidth and the derivative time are of importance. The proportional band would be a compromise between a small value, which may lead to instability and start-up overshoot problems, and a large value, which would give 'sloppy' control. The derivative time is usually fixed at about 1 min. This will give satisfactory performance with most machines, which will, of course, be within the limitations of this type of controller.

It should be noted that since the derivative action has a stabilising effect against system heat transfer lags, if the lag is very small (i.e. with a shallow thermocouple) instability may result. It is therefore advisable to use PD controllers only with deep thermocouples.

When using a three-term or PID controller it should be possible to achieve good start-up performance, good response around the normal working point, and no steady-state error, i.e. the PID controller should always control at set-point. Such controllers leave the factory with preset values of proportional band, integral and derivative time (typically 5%, 6 min and 1 min respectively) and these will give good performance on most machines (for zones with natural periods of less than 6 min). In general the response will be slightly overdamped so as to give some margin of stability (see Fig. 10). A small sacrifice will therefore be made in that the time for the machine to reach stable operating temperature will be a little longer than the absolute minimum.

There are however, cases where the 'standard' controller parameters are quite unsuitable and result in intolerable start-up performance or instability.

It would be possible to calculate the correct controller parameters by examining each zone of the machine in turn but the calculations are complicated and in any case unnecessary since there are a number of simple experimental methods which are commonly used.

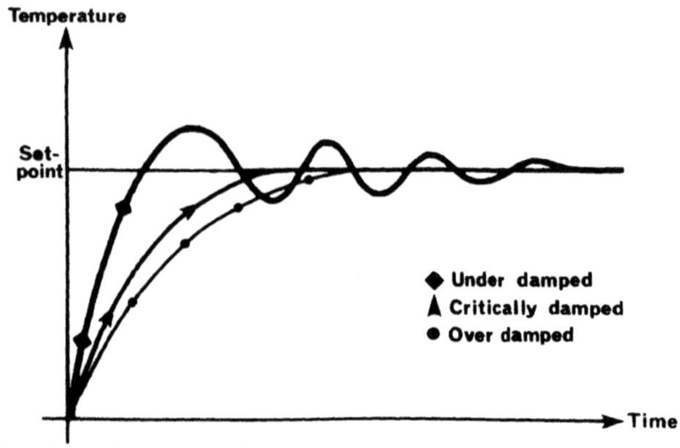

FIG. 10. Over-damped, under-damped and critically damped responses.

5.2 Experimental Methods

One method (Ziegler–Nicholls) is to remove the integral (by shorting out C_2 in Fig. 11) and the derivative terms (by open-circuiting C_1 in Fig. 11) and to narrow the proportional band until oscillation occurs as with an ON/OFF controller (see Fig. 4). The process is operated at normal temperature, and the natural period T_N and the peak-to-peak amplitude H_N are measured. An

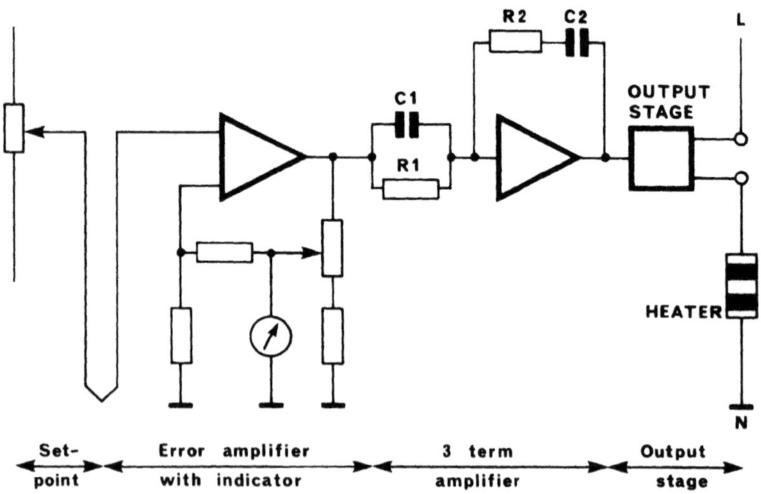

FIG. 11. Typical PID controller.

integral time constant T_I greater than or equal to T_N, is chosen and a derivative time constant T_D of $T_{N/5}$.

In the configuration shown in Fig. 11, the derivative and integral time constants are each a combination of the feedforward and feedback time constants T_f and T_b.

Thus, $T_f = C_1R_1$

$$T_b = C_2R_2$$

$$T_I = T_f + T_b = C_1R_1 + C_2R_2 \text{ sec}$$

$$T_D = \frac{T_fT_b}{T_f + T_b} = \frac{C_1R_1C_2R_2}{C_1R_1 + C_2R_2} \text{ sec}$$

Typical values are

$$C_1 = 2{\cdot}2\,\mu\text{F}$$

$$C_2 = 2{\cdot}2\,\mu\text{F}$$

$$R_1 = 33\,\text{M}\Omega$$

$$R_2 = 150\,\text{M}\Omega$$

Therefore

$$T_I = 2{\cdot}2 \times 10^{-6} \times 33 \times 10^6 + 2{\cdot}2 \times 10^{-6} \times 150 \times 10^6$$

$$= 402{\cdot}6 \text{ sec } (\simeq 6{\cdot}7 \text{ min})$$

$$T_D = \frac{2{\cdot}2 \times 10^{-6} \times 33 \times 10^6 \times 2{\cdot}2 \times 10^{-6} \times 150 \times 10^6}{2{\cdot}2 \times 10^6 \times 33 \times 10^6 + 2{\cdot}2 \times 10^{-6} \times 150 \times 10^6}$$

$$= 59{\cdot}5 \text{ sec } (\simeq 1 \text{ min})$$

Thus, if a natural period of 3 min was measured for a small machine then C_1 and C_2 would require to be about 1 μF for optimum control. The proportional band would be set a little wider than the peak-to-peak oscillation H_N.

After fitting the necessary components and adjusting the proportional band, the controller is allowed to stabilise at the normal set-point and the process is repeated with the next zone.

Another method[2] is particularly useful for cases in which the start-up performance is inadequate (e.g. too much overshoot, caused by a combination of high heater power—chosen to achieve rapid temperature rise, and deep thermocouple location—chosen so as to achieve close control of the melt temperature during normal running of the machine).

The procedure is to start with the machine cold and to apply full power (by setting the controller set-point higher than normal working

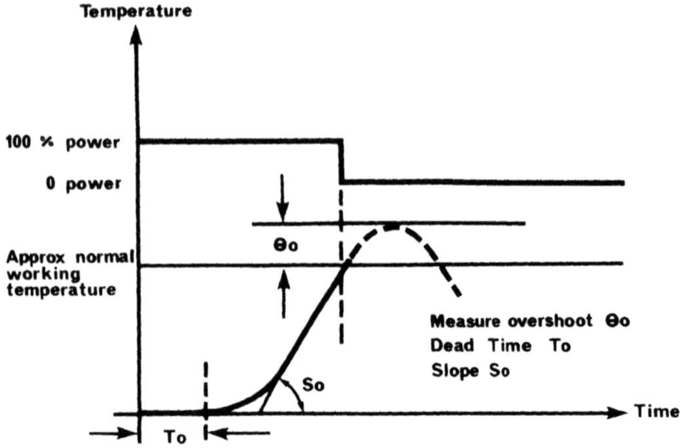

FIG. 12. Measurement of system parameters to determine optimum time constants.

temperature) until a temperature near normal working is reached, and then to switch off (Fig. 12). The overshoot θ_o in degrees Celsius, the dead time T_o in minutes and the slope S_o in degrees Celsius per minute are measured as shown in Fig. 12. A chart recorder is particularly useful for this purpose.

$$\text{Make } T_f \simeq 2 \times \frac{\theta_o}{S_o}$$

$$\text{Make } T_I\,(= T_f + T_b) \geq 1{\cdot}5T_o$$

T_I should be as small as possible within this constraint. T_f and T_b can then be realised in terms of resistor and capacitor valves to give C_1, R_1, C_2 and R_2. The overall gain should be kept as close to the original as possible—in this case

$$\frac{150}{33} + \frac{2{\cdot}2}{2{\cdot}2} \simeq 6$$

It should, however, be noted that most machines will be adequately controlled by equipment with standard parameters and only in extreme cases will techniques such as those described above be necessary.

6 HOT RUNNER CONTROL

Advances in cartridge heater technology have caused renewed interest in moulding tool design relying on the concept of hot runners. The use of small heaters capable of very high power densities in the mould has made it possible to predict very reliably the performance which will be achieved. The control of power is the usual method used in operating these heaters and the specification of a suitable controller would require a rating of 6 A at 240 V and a rejection of mains voltage changes of better than 10 to 1 (i.e. a $\pm 10\%$ change in supply voltage should produce less than a $\pm 2\%$ change in load power; for example, if the load power was normally 60% for a temperature of 180 °C, a 2% change in power would produce a maximum 3 or 4 °C change in temperature). Power controllers are usually rack-mounted, as shown in Fig. 8, to achieve high packing density. It should be noted that a 1 °C change in ambient would cause a 1 °C change in load temperature.

More recently, heater manufacturers have been able to include a thermocouple in the cartridge heater package which allows normal temperature control techniques as used in barrel control to be applied to the mould tool also. As in the case of the power controller, rack-mounting techniques are usually used but care must be taken to take account of the rather short time constants normally associated with such heaters.

7 CONCLUSION

The problems presented by the control of injection machine barrels and mould tools are now well-known and there exists a wealth of experience in companies specialising in these areas of temperature control. This can readily be brought to bear so as to provide an optimum trade-off between capital cost, reliability and performance, in particular cases.

Whilst the majority of temperature control problems are solved by purpose-built units of the type shown in Fig. 8, alternative solutions are becoming available. For example, microprocessor- or minicomputer-based systems can be used. However, the problems and the solutions offered are still basically the same, and it is only when full advantage can be taken of the increased capabilities of digitally based equipment that such systems will become an economically and technically viable alternative to discrete temperature control units.

BIBLIOGRAPHY

1. SOMERVILLE, M. J. *Eurotherm Technical Note* TN-6, System Controllability and 3-Term Control.
2. HALL, J. I. *Eurotherm Technical Note*—A Method of Setting Time Constants on Eurotherm Temperature Controllers.

Chapter 9

SOLID STATE SYSTEMS

F. COOKE

Technical Director, Optimum Evaluation and Instruction, London, UK

SUMMARY

Injection moulding machines are the victims of considerable prejudice. One does not hear comparable comments regarding extruders, blow moulders or vacuum-forming machines, but on nearly all the basic design features of an injection mould machine there are extreme views.

Examples are toggles versus hydraulic lock, hydraulic screw drive versus electric screw drive, hydraulic ejector versus mechanical ejector, and so on. The revealing word being versus.

The latest victim of this quite unreasonable and ill-founded attitude, because it assumes that machinery development stands still, is solid state control.

In considering developments the author seeks not to give a treatise on electronics, but to compare some items and systems, with which the old guard of relay enthusiasts are familiar, with the later and proven developments in solid state control that have now established their value—in short, to eliminate the use of 'relay versus solid state' but to recognise solid state as the successor to relay.

1 INTRODUCTION

It is a paradox that the plastics business, in general a modern, forward-looking business with significant developments on a yearly basis, should have been reluctant to accept comparable developments in the electronics section, for it is the transistor that has made solid-state controls such a flexible tool.

The progress of the solid-state controls in the plastics industry has followed the pattern of control in other industries, but has been deferred in application by about 10 to 15 years. The first applications were in temperature controls, as it was felt that this particular development could only be trusted for items that could be isolated and removed. Application in the early days was, in fact, resisted quite strenuously by the users, possibly to a greater extent than in any other comparable industry, and the early attempts sadly confirmed the view of the sceptics. The reasons have a great deal to do with the structure of the industry, particularly the moulding sector.

Skilled operators of injection moulding machines in many cases tend not to have electrical, mechanical or electronic knowledge, but just this type of individual with an almost animal sense of touch and 'feeling' can make a moulding machine produce good mouldings probably faster than his fellow. He then tends to gravitate to a senior position, still having no machine knowledge. It is understandable, therefore, that having 'learned' the peripheral skill of electrical maintenance there should be some resistance to accepting solid-state control where numbers appear in a neon tube and there is not even the noxious smell of a burning relay coil to give a clue to the area in which a difficulty may exist.

The style and logic used in the plastics industry and in particular solid state controls, are not always necessarily the most technically advanced components or the most elegant circuits, but they do reflect the manpower structure of the industry.

In general, therefore, the electronic components used have an accent on physical strength and circuits are used with a high degree of external indication of fault and cycle status that are peculiar to the plastics industry.

2 THE SOLID STATE

2.1 Definition

For the purposes of injection moulding control, the term solid state refers to transistor techniques. In computer development, electronic control and switching techniques have used thermionic valves and switches in glass envelopes as a stepping-stone to modern transistor electronic control. The same can be said of machine control by electronics.

The term 'solid state' is now the most widely accepted term, but it is felt that the less mellifluous phrase 'static switching' is the most accurate definition. The paradox of the basic element in this technique is that it is in fact neither a 'solid' (inferring that things will not pass, i.e. it is a perfect insulator) or a 'switch' (indicating that they will, i.e. that it will conduct). It is a region inbetween, now defined as a semi-conductor, with which we are concerned.

The switching technique is dependent on the extent to which the basic transistor material will vary between the two extremes of insulation and conductivity, and on the means of instructing it to do so.

2.2 Solid State Components

2.2.1 *The Diode*

The diode is one of the most easily understood components, and is analogous with a check valve in hydraulic circuits. A simple example of its use in machine control techniques would be where a control element, for example, a solenoid valve, has to be switched from more than one source at different times in the machine cycle.

With conventional electrical switching, a signal to this valve would also enliven all the other wires that are in common together with back-feed, thus causing circuit confusion. It would therefore be necessary to insert a protecting relay to prevent this at each circuit junction. The unreliability factor would therefore be increased. Diodes in each line would, on the other hand, allow common wiring, and prevent back-feed.

Diodes within the logic also fulfil this one-way protection requirement, and can be used as voltage regulators. Two basic constructions of diodes are used; they are (1) point contact diodes—these are usually devices in a glass envelope and are constructed using a silicon wafer, with a needle touching the silicon. This is a very fast operating device, principally used for high speed circuit logic, and (2) junction diodes—these have a larger contact area and can handle higher throughputs but more slowly.

2.2.2 *The Transistor*

This is another case in which an analogy is helpful. However, without going deeply into electron physics, it must be pointed out that for the purposes of a discussion relative to machine control only the major characteristics or features of the components (diode, transistors, etc.) are considered. For example, electron flows of values less than are used are possible in the opposite direction but are ignored.

The hydraulic analogy is with a very small pilot valve mounted on an enormous main valve. Very small changes in the pilot valve spool position can mean vast changes in the amount of liquid flowing through the main valve. The pilot control would in transistor terms be the base, with the through pressure connection lines being the emitter and the collector.

Basically banks of transistors can be encapsulated together in a way that would, in operational terms, be analogous with a relay having not only multiple contacts but also multiple coils. In other words, it can be triggered from several sources and, by using a gate, can also demand that a predetermined number of sequences are complete before switching.

2.2.3 *Transistor Gates*

It is possible to use this device to predetermine how many signals should arrive and in what order, so as to enable a subsequent sequence to take place. Such logic is inconceivable with a standard relay circuitry since, for example, to supply the coil for a simple control relay from several potential sources, requires each incoming signal to be separately wired through other contacts and relays. It is in this case abundantly clear that the advantages of the solid state logic are substantially greater than those of relay systems.

2.3 Circuits

2.3.1 *Circuit Concepts*

It is important to understand that the logic referred to is not specific to the electronics field. When considering system design, a useful method is to replace all the relays of an existing circuit that operate solenoids, with switching transistors.

The next step would be to protect this part of the system against solenoid failure; in a conventional a.c. system, this would mean fuses or trips. At this point an electronic device, having a set maximum value of current, can be employed; as attempts are made to exceed this level the device breaks the supply to that solenoid faster than any conventional fuse or circuit breaker (the emphasis being on the phrase 'as attempts'). The switching speed of the electronic system is of positive benefit within the logic but has little if any effect on the overall cycle. Conversely, should the solenoid go 'open-circuit', it is reasonable that this should also be monitored and indicated.

A written description of solid state current operation, when compared with actual values read on a meter, can be confusing. To the fault-finding engineer familiar with standard electrical systems, who would normally connect one side of the test meter to a neutral or grounded connection and then expect to read line voltage via the other probe, a new concept must be taught. In the standard system an a.c. voltage (110 or 240 V) passes round the information within the system either by its presence or by its absence; in other words, either zero or the line voltage may be read. This is not usually the case with the type of logic used on injection moulding machines.

It is important to note that the signals passed around in solid state systems are usually referred to in the machine manuals as '0' or '1'. This does not refer to specific voltages, but to the difference that the system designer has elected to have between his 'go' and 'no go' signals. The exact value is usually defined on the drawing. Again an analogy would be helpful. Consider an oil pressure of 100 psi. In solid state logic this would compare with 100 psi being the '0' and 150 psi being the 1, but it can be seen that although '0' is used it is not zero, nor is 150 psi equal to '1'. Very few electronic systems use zero volts as a reference voltage.

A minor mental adjustment, together with a test meter re-calibration, is called for, since it can be seen that a system can be switched off logically ('0') but can be reading voltage.

There is a range of controls for machines that have a built-in meter, which, to avoid the confusion of multiple readings, just shows a light when a component under test is in working order.

2.3.2 *A Machine Circuit*

It will be seen that the function plan (Fig. 1) is divided into sequences. Before considering this it should be remembered that any order from the control to the machine requires an acknowledgement from the machine when this has completed the given order. In other words, after each completed sequence, the logic unit must have an acknowledgement, maybe not only from the machine but also from the safety circuit, the timer unit or the programme unit, before it gives the order to the machine to carry out the next operating movement.

We start at the extreme left of the function plan (seq. 1). The first symbol

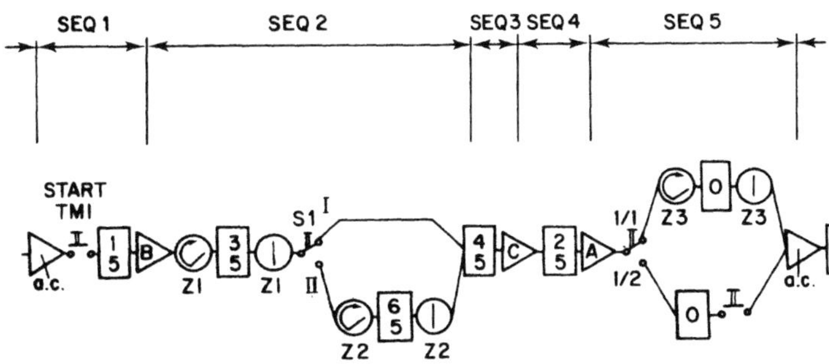

FIG. 1. Function plan of a machine circuit.

indicates that a condition for starting is that the limit switch marked a.c. is activated, e.g. by closing the machine guard. By pushing the button TM1 (start) down, the output amplifiers will supply a voltage of 24 V d.c. to the solenoids marked 1 and 5. The solenoid will respond, and the machine will start the closing movement and continue until the limit switch B is mechanically activated. The electrical signal from B is sent as an acknowledgement for the completed movement back to the control.

The logic unit now gives two simultaneous orders. It orders (1) the start of timer Z1, which in this case is pre-injection time, and (2) the output amplifiers to send power to the solenoids 3 and 5 (high pressure). The machine injects. As seq. 2 is not determined by a distance but by a timed movement, the timer Z1 takes over the acknowledgement of the completed movement when the set time is up.

At this place in the function plan, we see the symbol for a program key S1. In other words, at this point we can choose between two programs, 1 or 11, for the further course of the machine function. If we have chosen program 11, the timer Z2 will start; at the same time the solenoids 6 and 5 (low pressure) will operate, and the machine will now continue low-pressure injection until the timer Z2 acknowledges the completed order when the time is up.

Now the solenoids 4 and 5 will operate and the machine starts plasticising. This continues until the adjustable limit switch C is mechanically activated and remains activated until the next injection. The limit switch C takes care of the acknowledgement of the completed movement. Solenoids 2 and 5 operate and the machine opens up and goes back to the limit switch A, which acknowledges.

Here again we meet a program key, e.g. semi- or fully automatic operation. We choose fully automatic and see that the timer Z3 starts. At the following symbol we can read that no solenoid is operating while the timer Z3 is running. When the timing of Z3 is finished the control orders the start of a new cycle.

2.4 The Advantages of Solid State

The advantages of replacing the relays (used in conventional systems as back-feed preventers and inhibitors) are considerable. Relays and back-feed interlocks are replaced by diodes, and well-proven electronic switching techniques replace the rest of the components. This is a gross simplification; however, it can be seen that using the replacement technique for redesigning machine control systems means that the electronic system has many inter-circuit or component advantages that take up very little space but offer benefits adding a degree of reliability and performance unheard of in a.c. switching systems.

The solid state components are nowadays mounted on cards. The cards are boards with a high insulation value and of a thickness ranging between 0·5 and 3 mm, faced with a lacquer that will accept the other advantage of solid state technology, the printed circuit.

The advantages of the replacement of the wires in a machine will gladden the heart of most electricians. Otherwise, while standing in open-mouthed amazement at solid state technology, the transistor system would appear to be an impenetrable mass of live wires, ready to harden and desolder, and so keep their defects hidden until only the most skilled could detect their failings.

To print the interconnection by etching methods means that the small components can be conveniently located close to each other with no mechanical joints and, if the manufacturer does his job well, the edge of the card can have an edge connection enabling the card to be inserted as a unit into a matching contact block. It is common practice for the blocks to be racked in compact rows.

A traditional connection switching system for a 120-ton moulding machine has, on recent examination, approximately 2700 screwed connections. The replacement solid state control was found to have 230 such connections. It would be unwise, therefore, to ascribe all the benefits of solid state controls to the transistor, since the elimination of so many potential loose screws and hardened wires, as well as failed insulators and mechanical joints, must be of enormous and often ignored benefit.

So confident is one company of the advantages and reliability of its logic and components that it guarantees its system for 5 years, a degree of reliability that was unheard of even 5 years ago.

It has recently been established that in one case, where a machine manufacturer now supplies solid state controls as standard, the call-outs for service of an electrical nature have fallen to 18 % of that experienced with conventional switched machines of the same type. Two conclusions can be drawn, namely that on solid state controls the machine user is able to identify his problems and/or that they are now less frequent.

3 BASIC DEVICES

Before a lengthy discourse on injection moulding controls is made, some discussion is necessary. Whatever the machine sequence system is, be it relay or solid state, the controls must check, for example, that sequences are proceeding in the correct order and that go and no go signal devices are working. In such cases feedback can be purely an internal happening in a series of components, without an external indication that it has occurred; however, it is conceivable that if such information is not forthcoming an external signal can be produced to show that a pre-condition of machine sequential safety has not been met.

To enable very simple external indications to be made that the machine system is working properly, external wiring, and in some cases components, must be added, the net result of which is the physical enlargement of the overall system.

3.1 Transistor Assemblies

With the introduction of transistor 'chips' it is possible to have, within a small space, the function of several transistors the output of which controls one function with potential inputs from several sources. As mentioned previously (Section 2.4), it is now common practice to have the control circuits for all the solenoids isolated on a single board for ease in fault-finding, and often the boards have identical components. This is made a practical reality because of the development of such 'chips'.

In some cases the solid state control cards that replace relays are, for example, physically stronger than they might need to be, since they have to be handled by inexperienced staff.

3.2 Limit Switches

The limit switch has always been a major source of indication of function arrival or setting. This was one area that did contribute to the problems that were attributed to solid state control. Attracted by the fact that transistor circuits could operate on very low voltages, early solid state moulding machines were produced that used their original switchgear but used 6 or 12 V. The results were disastrous. After a while the switch that had been designed for higher voltage suffered from self-insulating contact points. Gaps across which 220 V a.c. would leap, were too much for the lower d.c. pressure and, in several cases, d.c. proximity switches were physically too lightweight to suffer the ravages of conventional and hurried mould-shop treatment.

It is ironic that the application of solid state controls on moulding machines was held up for the want of switchgear with the suitable mechanical function and not because of the solid state itself.

3.3 Relay Replacement

One area where the reliability of solid state control is showing substantial advantage is where relays have been replaced by transistor components. The general failure within relay logic systems is either coil failure or coil burn-out. Although relays have become one of the most highly developed of the electro-mechanical devices, external forces can affect them. A typical example is where an a.c. hydraulic solenoid is working up to its design limit, switched by a contact on the relay. Repeated switching can erode the contact points. To use a larger relay would mean that reaction times would be slower. This is a classic takeover situation for a solid state switch that would save time. In addition, very simple circuits could be introduced that, in conjunction with a d.c. solenoid, could automatically register a malfunction should the current in the solenoid either be zero or be an agreed amount above a norm.

3.4 The Timer

A very common application of solid state control was originally the use of a resistor/capacitor combination as a timer. This was basically a method of timing that gave an alternative to the complicated mechanical devices which had been the mainstay of the control of machine sequences. The principle of this solid state device was that a variable resistor metered power into a container (the capacitor) which, when full, discharged its contents so as to operate a small relay which could then switch major functions. It can be seen what an unhappy marriage of techniques this is, since there is the following chain of events: mains a.c. into low voltage (transformer) into d.c. or partial d.c. (rectified current) through a resistor into a capacitor, then a d.c. relay switching a.c. current.

Most capacitors, as used in the earlier timers, also suffered badly from the effects of dampness and ambient temperature variation. The resistor/capacitor was therefore an unreliable device when high speed accurate cycling was required. In general, this hybrid has been relegated to less sensitive areas of application, e.g. air timer blasts sensing circuits reaction, i.e. where the actual measured time can stand a percentage of error. It must also be remembered that the times set on the dials of such devices are inaccurate and, when running, no moving parts can be observed. Because of this, external indicator lights are used. Such devices have two characteristics to recommend them, namely that they are compact and they are cheap.

The timing of sequence is where great accuracy is demanded. Solid state techniques are also being applied here but, oddly enough, using one of the main operational principles of the older 'clockwork' timer. Whereas mechanical timers use a synchronous motor—in other words, one which is mains-frequency related—in general, solid state controls use a system of counting a.c. pulses in order to give a time base. Again, external indication must be given.

It is not uncommon for the timing unit of a machine to have one central mains counting device where the incoming supply a.c. pulse is converted from the sine wave form into the electronically more desirable and more intelligible square wave form.

For mechanical convenience, rows of thumbwheel digit wheels are banked together in front of the basic counting unit. A number of pulses is set and at their successful conclusion a signal is passed on.

This type of layout enables ready interchangeability of electronic and mechanical components; again, components on edge-connecting cards ensure simple replacement.

3.5 Solenoid Operation

The most common method of solenoid valve operation had used alternating current as a matter of convenience, usually at 220 V. The mechanical structure of such solenoids is that a laminated core is displaced from a 'rest' condition. When current is supplied, the windings surrounding the core are energised; a massive inrush of power takes place until the hammer (armature) goes home. This inrush is the enemy of the system components, since, if held for too long (e.g. by spool jamming, excessive oil flow, or by over-frequent use), it readily burns out. This is because the individual wire strands used in the manufacture cannot stand the current and the insulation deteriorates as the valve becomes hot.

In general, direct current does not suffer from the disadvantage of the inrush current being so much higher than the hold current. In addition the voltage required is less. Under the right conditions, therefore, d.c. solenoids are desirable but relatively expensive.

Used in conjunction with solid state switches, it is not unknown for d.c. solenoids not to be replaced throughout the life of the machine. Where the solid state circuit incorporates circuits to protect and inform on the condition of the solenoids, care should be exercised when adding solenoids for the operation of ancillary circuits, e.g. for core pulling, etc. This is one of the few areas where solid state system is less flexible than the conventional relay operation, unless ancillary stations have been pre-wired into the system.

4 APPLICATION OF SOLID STATE

4.1 Safety

Devices which indicate oil level, oil temperature and oil filter condition all pre-date the use of solid state control but they have tended to become standard because of the simplicity of incorporating warning systems into the logic. The subject of filter condition has also another significance that is described later.

With some confidence it can be stated that whilst solid state control has not added in any way to the practical side of this subject, it has made the detection of faults within these systems much easier.

4.2 Temperature Controls

There are in existence many hundreds of temperature controls that used solid state controls as the basis of measurement. In one such system output from a galvanometer was used to deflect a needle. A flag fitted to the needle interrupted a photocell beam to produce current change. The photocell, usually a OCP71 photo-transistor, has its signal amplified—by a thermionic (valve) amplifier—sufficiently to energise a relay whose function was to switch on the heater band.

It was at least ten years after the introduction of this partial solid state device, that solid state temperature controls became universally accepted. They did so in advance of machine solid state controls, because, to the moulders, they were new but could be removed and replaced if found to be troublesome.

Temperature controls using solid state logic are now universally accepted and two refinements are worth noting, but the difference between using solid state logic and solid state switching is particularly important. The former does not necessarily include the latter. Simple and effective temperature controls use the input from a thermocouple, compare it with a variable value set on a dial and switch a small relay on or off.

In an effort to control the overshoot, bearing in mind that there is an electrical inertia in the heater bands, anticipatory devices were introduced that momentarily switched off the controller on its way up to set temperature when starting from cold.

Well proven electronic techniques have been applied that allow switch off to occur even earlier should the rate of climb be too fast, i.e. when heating an empty material cylinder; other variables (such as drift above and below set point and frequency of on/off switching) can be changed at will.

The design of temperature control logic is now so well and equally developed that it is now the question of external design and display that exercises the mind of the user, other features being equal; however, digital, thumbwheel setting is the most noteworthy change.

Having dealt briefly with logic, and the way it is used to operate on/off relays, the matter of power switching can be considered. There is some evidence to suggest that switching heater bands on and off at mains voltage can be less effective than keeping them switched on but at reduced output.

The power transistor is a natural for this duty. It can protect the heater bands in two ways. Firstly, when starting from cold, when the curve of time and element temperature would be steepest, it can apply a reduced supply to 'dry out' the element. Then, instead of switching on and off, it can apply an increasing amount of electricity to the band which can benefit from all the anticipatory logic in the control unit without switching off the power, i.e. it can be controlling the electrical flow to what is required.

The siting of this equipment in a control cabinet should be carefully arranged, since power transistors get very hot. Ventilation is important since an unreasonable temperature rise in a control cabinet could unbalance the machine logic.

4.3 Machine Setting

Line-signalling devices as applied to injection moulding machines have met with varying degrees of success. Consider the standard machine, whether it be toggle or hydraulic lock. It has always been possible to digitally set the distances and positions of sequences (i.e. mould opening, closing, clamping and ejector operations), but the most important requirement is that such equipment should be reasonably economic. It is not at all clear that remote setting using linear signalling devices has financial benefits.

A problem with moulds is that they can have a varying thickness as a result of temperature changes, and alterations in oil temperature can affect closing and opening speeds. Also the measuring equipment that is required and is available (e.g. linear displacement transducers) does not stand up to the very real possibility of mechanical damage.

All the foregoing objections can be overcome electronically by, for example, compensation for mould thickness variation, oil temperature, etc., but the expenditure and complications would not be countenanced by the average moulder.

Punch-card systems have been applied to the switching of core pulling sequences but these were of the crudest type and relied on microswitch fingers which bridged through the punch holes. Although this was an ideal method of repeat setting of complicated core pulling programmes, the mechanical devices of fingers and probes were less than satisfactory.

The application of existing computer peripheral equipment to pre-programming also has a dubious record. This is in spite of the attractive possibility of external links to a control centre giving indications of production rates and machine performance.

Machines have been constructed with tape input, pre-setting and central information correlation, but if the normal moulding shop conditions are considered and compared with the delicacy of a standard punched tape or card, the reasons for their failure are obvious. Information on punched data would need to be indented on to steel cards to have any life at all.

It must not be assumed that for these practical reasons such refined equipment has to be discarded, and that there is some basic lack of intelligence in the industry; poor application of the computer software by those more used to office equipment, is probably the answer.

4.4 Cavity Pressure Control

Conventional electronic components have been adapted for use on injection moulding machines but with varying degrees of success. Cavity pressure control has occupied the time and endeavours of plastics engineers; it has two possible benefits and is applied in several ways.

It is generally accepted that the pressure within a mould during the filling (injection phase), has only a fleeting relationship with the pressures shown on an injection oil pressure gauge. This is basically because of the flow pattern of the material as it is pushed through orifices of varying size (e.g. sprues, runners, gates) and over surfaces of varying temperature.

Pressure stepping the injection feed by limit switch control has proved to be an economical method of approximating the ideal and refinements such as speed stepping has also been reasonably effective. On relay machines, as many as six steps have been tried on injection pressure but the proximity of the limit switch can be badly affected by external factors, for example, material feed rate, fluctuations in material and mould temperature, etc. It was also thought that the locking forces on machines could be reduced, if the injection pressure could in some way be reduced to a low level, consistent with a sink-free moulding, after using a pressure that adequately filled the mould. The developments that have taken place have only been possible as a result of the application of solid state control.

Cavity pressure control has been only marginally successful in the area of locking force reduction, but in the area of production quality and density control it has had resounding success. One of the electronic sensing devices that has made this possible is the pressure transducer. This comprises two or more discs of piezo-electric quartz with a conducting material between them; to this conductor wires are attached. Another device is based on a semi-conductor material whose resistance varies with pressure (rather in the same way that temperature affects thermistors). In the former the piezo-electric device operates by electronic vibration and in the latter by variable resistance and the effect brought about by pressure.

A change of state caused by stress is readily measured and amplified so as to operate a solid state switch. However, where, for example, a machine is of the relay type, it could introduce, with some significance, a second-stage pressure signal. Wide operating temperatures are possible. It is essential that this device is positioned where the effect of pressure change is of some importance. In cases where an existing mould is considered, a modification to the ejector pin system can be effective.

A simple way of establishing the best place to place the sensor is to produce the moulding, and then reduce the pressure in small steps. Where the moulding starts to 'short' is the area that is the potential site.

4.5 Feedback and Servo Systems

4.5.1 *Feedback*

The subject of solid state control cannot usefully be discussed further without some definition of terms being made. 'Feedback' is probably a word that is as commonly misapplied and as misleading as the phrase 'major breakthrough' in public relations. The very use of the term in machine control systems tends to imply that merely having information about a condition or change of state is adequate. The term is unconsciously applied on the internal logic of a machine, for example, where transistors provide signals only when a previous signal has been responded to, or where a solenoid's electrical consumption is acknowledged as being within limits. Such events take place in microseconds, but they respond and react. Feedback, or information, is only as useful as the application to which it is applied.

The machine's basic hydraulic drive is, without doubt, a major factor in determining the type of control used and, for reasons that relate to field maintenance and component supply, fixed delivery pumps, in combination or with accumulators, are the legacy of the injection moulding machines.

This latter point is a typical case where a technically elegant design solution, i.e. a marriage of solid state control to infinitely variable volume- and pressure-compensated pumps, is in practice unsuitable and the real-life consideration of moulding machine maintenance must apply.

Accepting, therefore, that the oil that has to power the machine is available either as a single volumetric output or stepped as functions of the multiple pump assembly, the external control must rest in the valve gear. It is a simple matter to instruct a valve but what is the value of its response or feedback? This question is the subject of current debate.

4.5.2 *Servo Equipment*

Cavity pressure control is a case in which the use of a signal from the cavity can conceivably alter a servo relief valve head setting. Compared with a known value (set in a digital electronic control), and adjusted accordingly, this would be a 'closed loop' feedback system. An 'open loop' would be where the digital valve is set but no reaction is applied. In cases of machines cycling at 2 s the hydraulics cannot respond in time and it has been shown that, so long as the oil viscosity is stabilised at the high injection pressures and speeds used, the expense of closed loops is a luxury.

Servo equipment which can vary speed is now frequently fitted to injection moulding machines but, in general, such devices are limited to the injection end of the machine. This is yet again a problem of solid state control with digital setting, as the erroneous setting of a digital switch could produce catastrophic results on a locking unit; on the injection end, however, such losses would be recoverable.

Buhl, Hunkar and Dynisco have long been in a position to provide open loop systems of injection stroke and pressure control as 'bolt-on' units. They have been either peg-board, digital or analogue systems with sliders. Development and application have without doubt been delayed for the want of appropriate non-filtration-conscious servo valves, which now appear to be coming on the market.

In the case of older machines and solid state machines that lack the facility, either servo valves or banked digitally set relief valves must be fitted. This is normally an expensive procedure as one prime requirement of servo control is the introduction of oil that has been filtered to less than 10 μm. This should be compared with the acceptable norm of 20–25 μm. The introduction of the clean oil system can be extraordinarily expensive when coupled with the cost of the servo valve (whose physical fitting is simple). Although an item of hydraulic equipment, torsional servo valves tend to retain the price aura of their parents, aircraft hydraulics, rather than that of the less sophisticated poor relative, the plastics industry.

In general, cavity pressure control as a simple solid state facility, tends to remain a pressure-step adaptation. Some very important work has been done on cavity pressure control at the National College of Rubber Technology in the UK, using a Josef Krauss KHS40, a 40-ton lock machine.

Coupled with the availability of servo equipment, linear transducers have also made possible closed loop hydraulics. They do, however, in many cases tend not to stand up to the rigours of the plastics industry, and where this equipment is now being used substantial guarding is the order of the day. Pressure setting is becoming a control panel function and, again, the piezo-electric transducer comes into its own.

A fairly recent application of servo solid state driven equipment is that of the screw speed. This is probably a case where the actual values of response are not important and an open loop with an adequate indicated feedback would be suitable. The reason for its use is based on evidence which suggests that there is some benefit that can accrue from the screw's rotating at high speed initially but decreasing as the melt volume increases, therefore keeping the plasticised material at a constant temperature throughout its axial section.

5 SOME TYPICAL SYSTEMS

5.1 Rack Systems

The accessibility of the cards and the racks into which they fit is of course important, and one of the grey areas of the application of card systems is the number of cards and how they can be tested.

Many of the components that are repeated are the logic sequences. An early attempt at controlling a machine by solid state had all the components on two large boards; this was a fault-finding nightmare. Although isolation of components into groups is more expensive than mass boarding, it pays dividends in maintenance and subsequent modification when required. It is equally foolish to go to the extreme of each step being isolated on a single board. (There is a machine that uses more than 21 different cards.)

By using segregation techniques it has been shown that a machine can have up to four different types of card and can, for example, use a total of twelve. It means that within some cards there will be unused portions when used in particular situations but for test purposes the cards can be exchanged within the machine.

5.2 Application and Development

It is not unreasonable to suggest that within 2 years the relay machine will only be sold at a premium price. There are already several companies that sell only solid state machines, and one that charges a premium to supply relay systems.

Arburg have for several years been producing their own solid state controls for a limited range of machines with commendable results. For the others, however, there is a problem looming on the horizon. As marketing pressures start to be felt by the machine builders, there is a danger that they will all seek their own route to the promised land of solid state controls.

The desirable solution is that, in the same way as for hydraulic components, supply is in the hands of experts who retain their own service network for solid state. Three companies have risen to the challenge so far, General Electric, Buhl and Schlecher, and the market is so vast that others will follow. They all offer racked systems with fault-finding methods ranging from direct light-emitting diode switch indicators and test positions for suspect cards, through to pre-wired integral test meters. All of these devices are inappropriate to relay controls.

6 CONCLUSIONS

Solid state controls are now an established fact of injection moulding life. The application of the better systems has brought about the first significant development in injection moulding machines since the application of the screw. It cannot be ignored since the intelligent user will become more profitable than the user who adheres to relays.

Computer techniques were applied to injection moulding machines too early in the development of solid state controls, but over the next few years there are real possibilities. One obvious area of investigation will be the solid state control of all the plasticising functions, as a result of the application of thyristor motor control, and the possible elimination of hydraulic screw drive. There is at the moment a real possibility of the electrical screw drive being variably speed-controlled as a result of information supplied from the melt. The variables which could be controlled include melt temperature band, melt and cavity pressures, injection and screw speeds, and screw back-pressure.

Existing techniques in solid state controls, coupled with computer techniques, could eliminate many setting functions on the injection unit. The industry would do well to observe the developments in variable pump and motor systems currently being made on numerical machine tools.

Chapter 10

PROCESS CONTROL

G. F. TURNBULL
Joint Managing Director,
Turnbull Control Systems Ltd, Worthing, UK

SUMMARY

General principles of more traditional process control industries are increasingly being applied to the injection moulding manufacture process. In Section 1 *of this chapter, recent developments in process control, particularly with regard to the move towards distributed intelligence, are outlined. The application of these ideas to injection moulding is examined in Section* 2 *and an overall block diagram of machine measurement and control units communicating to a central unit is described. The three main control units—temperature controller, velocity/pressure profile controller and programmable logic controller—are examined in Sections* 3, 4 *and* 5. *Section* 6 *describes an alarm annunciator unit and a production monitor unit is described in Section* 7. *Section* 8 *discusses the handling of input signals and Section* 9 *the power control units for the main hydraulic and electric circuits. Having looked at the component parts in detail, the overall machine monitor and control unit and its computer interface are outlined in Section* 10. *Overall production control for a complete moulding shop is discussed in Section* 11 *and some advanced control strategies are briefly discussed. Finally, in Section* 12, *there are some general considerations of how things may develop in the future.*

1 RECENT DEVELOPMENTS IN OVERALL STRATEGY FOR PROCESS CONTROL

1.1 The General Problem

Figure 1 is a simple illustration of the general problem confronting the process control engineer. On the left are the various analogue and digital signals necessary to monitor and control the process accurately and on the right are the output devices by which the process may be controlled. These are linked by control, monitoring and alarm devices communicating also with an operator's panel.

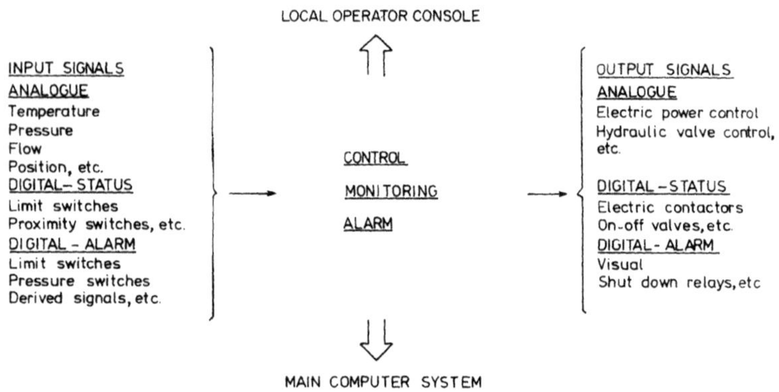

FIG. 1. Process control—the general problem.

1.2 Conventional Approach

Using electronic instrumentation, the conventional (often referred to as analogue) approach to monitoring and controlling a process is:

(i) to decide which of the analogue variables can be put under closed loop control;

(ii) to implement this closed loop control by a unique controller (temperature controller, flow controller, etc.) with the relevant transducer at its input and an electrical signal of the right form to drive the appropriate power element as its output; these controllers would often contain alarm circuits to detect dangerous conditions, and special features such as overshoot inhibition;

(iii) to decide which of the analogue variables are required only for monitoring and to connect these, possibly via switch units, to dedicated indicating devices, recorders and printers. Again, alarm circuits are often integrated into these devices;

(iv) to decide which of the digital signals are necessary for control, either in a sequential or in a consequential manner;

(v) to implement this control by hard-wired, dedicated logic using either relays or solid state equivalents;

(vi) to decide which of the digital signals are necessary for alarm purposes (some of these would originate from controllers and indicators as explained);

(vii) to interconnect these with hard-wired logic to give appropriate indication and often to facilitate shut-down procedures during fault conditions and provide indication for fault diagnosis and repair.

1.3 Computer System Approach

With the advent of the computer it became possible to implement closed loop control, analogue and digital monitoring, and consequential and sequential control and alarm strategies, all in a single computer. In fact many large computer schemes have been implemented in widely differing manufacturing industries. Obviously, one of the main considerations when assessing a computer solution versus the conventional approach is economics and one would therefore expect (with the advent of larger scale integration dramatically reducing the size and cost and greatly increasing the power of the mini-computer) that the computer solution would become more and more widely used. Whilst, in general terms, this is true, the way in which mini-computers are tending to be used and also the way in which microprocessors are being much more widely applied need looking at in some detail.

1.4 Distributed Intelligence Approach

There are certain basic advantages in using a single computer to do a multitude of tasks such as controlling a large process or several smaller processes. One problem is the large initial capital outlay which may be justifiable on a new installation but which presents difficulties when the desire is to upgrade existing installations. Another fundamental problem is that a malfunction in the computer can have very severe consequences. Both these problems are overcome by distributing the intelligence. This is in some ways similar to the conventional approach but uses digital techniques in the implementation of the measuring devices, control loops, etc. Put simply, a system is divided into a number of component parts using criteria such as physical location, plant security, plant operation, etc. Each of these component parts has then built into it a large degree of intelligence. It has become possible to do this economically because of the increasing power and decreasing price of the microprocessor.

Despite similarities between the conventional approach outlined in Section 1.2 and the use of distributed intelligence, there are significant differences as follows:

(i) More complex functions can be readily implemented using the microprocessor.
(ii) The use of fewer components can lead to greater inherent reliability.
(iii) Local self-diagnostics can more easily be built in, allowing malfunctions to be quickly detected.
(iv) It is possible for each device to put information on to a central highway in order to communicate information to a central mini-computer. The main functions of this central mini-computer would be data collection, production control and report generation. In addition, it may perform higher level control strategies—these, however, being such that the plant would operate satisfactorily even if the mini-computer breaks down.
(v) Also, rather than dividing a process into individual loops, it may be sub-divided quite differently, e.g. into a group of input variables, a multi-loop control package, etc. This will become clearer in later sections.

1.5 Communication Bus Structures

As explained in Section 1.4, to make optimum use of a system using distributed intelligence, it is necessary to have a communication network or bus structure whereby devices can communicate with one another and also with a central computer. Many systems have been evolved but perhaps the most sensible structure for the types of system to be discussed later is that shown in Fig. 2. On each section of the plant, devices having a significant degree of intelligence communicate on to a parallel highway (multicore cable). This can offer significant simplifications and higher speed over a serial bus (twisted pair of wires), but is restricted to short distances. It is therefore advantageous to use interface units in order to communicate on to the main serial highway. The computer is also connected on to this serial highway by another interface unit and usually takes command. More complex structures require units on the highway, generally referred to as highway directors, which determine, for example, signal priority.

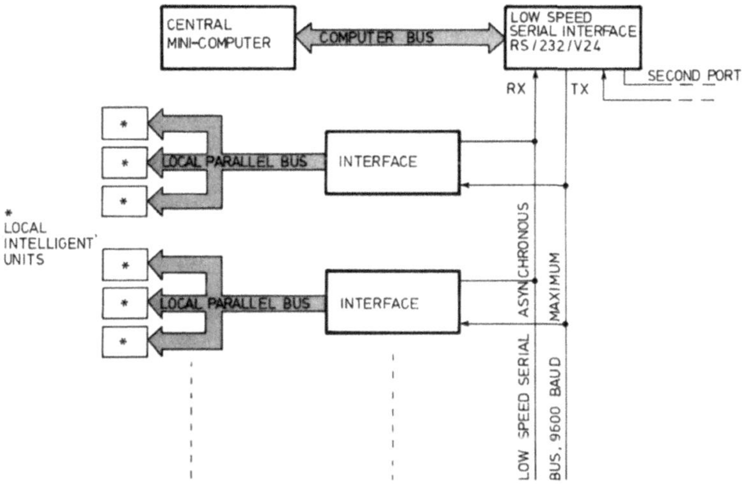

FIG. 2. Example of bus structure.

Over the last few years, certain standards have been generated for communication buses. These are as follows.

(i) 16-*Bit high-speed parallel bus* (IEEE 488): this was originally an International Electrotechnical Commission (IEC) standard adopted by the Institute of Electrical and Electronic Engineers (IEEE). It is implemented by Hewlett-Packard as HPIB (or GPIB) and has become widely accepted of late. Eight lines are used for the bi-directional flow of data, three lines for timing and control and the remaining five lines for interface management purposes. It is mostly used for measuring instruments and has a possibly restricted application in the context of injection machines (see later).

(ii) *Low-speed serial asynchronous bus* (CCITTV24RS232/V24): this is an international telecommunications standard implemented in America as RS232 and allows very simple low-cost cabling with twisted pairs. Any computer with a 'teletype' interface can drive the link directly at up to 9600 baud (one bit per second). It is highly relevant to injection moulding (see later). If this bus is too slow, either a standard high-speed synchronous serial interface or multiple RS232 buses can be used. There are many standards for high-speed serial links such as the IBM system widely adopted as SDLC (serial data link communication), working at 48 000 bauds.

The problem with these links is the increased protocol (error indication routines, etc.) and, consequently, the higher expense of the associated equipment.

The application of distributed buses and associated highways will now be considered in the context of injection moulding.

2 APPLICATION OF THE DISTRIBUTED INTELLIGENCE APPROACH TO INJECTION MOULDING

2.1 The Basic Approach

The basic concept of distributed intelligence applied to an injection moulding factory is illustrated in Fig. 3. Each machine is equipped with a monitoring and control unit. Communication to a central mini-computer is via a twisted-pair highway as described in the previous section. The monitoring and control unit on each machine is further sub-divided into units possessing significant intelligence and communicating via a parallel

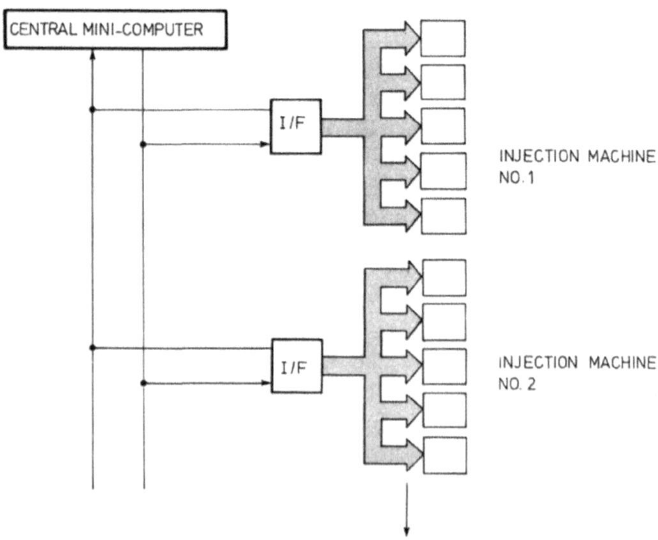

FIG. 3. Bus structure in an injection moulding factory.

bus similar to that described in the previous section. Before proceeding to examine the exact form which the units might take, it is useful to look at the fundamental nature of the injection machine with particular regard to monitoring and control.

2.2 The Injection Machine

It is conventional to divide the machine into two main units: (i) the mould unit, comprising fixed and moving halves of the tool; and (ii) the injection unit, comprising barrel, nozzle, screw and hopper. The injection moulding

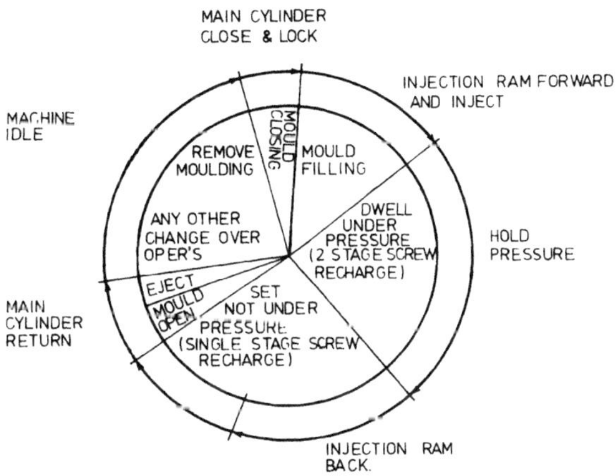

FIG. 4. Injection cycle

process is basically a cyclic one. The main elements of the cycle are illustrated in Fig. 4, showing the main sequence for the mould unit and the injection unit. The first fundamental is to control this sequence. Let us, for the purposes of this discussion, assume that the mould unit control and injection unit control are carried out separately. This is often convenient in order to make it easy to change mould units from machine to machine but in addition there is a very strong case for fitting a position transducer to the injection unit, whereas this is less easily justified for the mould unit (as discussed later). A unit controlling velocity profile versus ram displacement, hold-on pressure versus time and back pressure versus displacement would then be used for controlling the injection unit (see Section 4) and a programmable logic unit (see Section 5) could then be used for the mould control. Pressures and flows for the hydraulic circuits are all controlled from these two units in accordance with the injection cycle.

Before injection can take place, the temperatures at the various parts of the injection and mould units must be controlled. Almost all temperature control zones are on the injection unit so temperature control of the fixed and moving halves of the tool will be considered together with that for the injection unit (Section 3).

Large forces and high temperatures are features of the injection unit which make it essential to have comprehensive alarm and protection features. Alarm signals are derived from transducers fitted just for this purpose or by detecting excessive values of variables under continuous measurement or control. It is extremely unlikely and contrary to the principle of distributed intelligence that essential alarms and shut-down procedures would be performed in the central mini-computer. Therefore, a separate alarm annunciator would be fitted on each machine (Section 6).

In order to effect proper production control it is also necessary to monitor certain parameters, some of which are necessary for control and others which have to be provided in addition (this is discussed in Section 7).

Thus we have defined the five main units for each machine: (i) temperature control unit; (ii) velocity/pressure profile unit; (iii) programmable logic unit; (iv) alarm annunciator unit; and (v) production monitoring unit. These will now be discussed in detail with regard to their practical implementation. It will then be possible to define what inputs are necessary and how these are best handled (Section 8) and also how they interface with the power control units for the main electrical and hydraulic circuits (Section 9). Finally, the overall machine monitoring and control unit with its computer interface will be examined (Section 10).

2.3 Operator Interface

For the purposes of the next five sections describing the five basic units, it will be assumed that:

(i) all important analogue set-points will be on digit switches;
(ii) displays will be analogue for control loops (in the form of deviation displays) and digital for some key parameters;
(iii) all machine function selectors will be rotary switches;
(iv) all initiation-type controls will be pushbuttons.

Alternative methods of setting and display will be discussed when the machine measurement and control unit is discussed as a whole in Section 10.

3 TEMPERATURE CONTROL

3.1 Basic Principles

The techniques of temperature control are discussed in detail elsewhere (see Chapter 8). Therefore, only those additional features required for central monitoring and control will be covered here. In this section we are concerned with implementing closed loop control. As explained, this has been done traditionally with individual analogue networks and this could still be the best approach in some areas. Use of digital techniques can, however, offer the following advantages:

(i) Cost: with a large number of zones, a microprocessor will soon, if not already, offer a cheaper solution.
(ii) Modification of control terms: optimum three term settings will obviously differ between some barrel and nozzle zones but changing these settings according to machine parameters will, in most cases, give only marginal benefits.
(iii) More complex algorithms: start-up, overshoot inhibition, power feedback, removal of non-linearities. Although much can be achieved with analogue circuits, it is almost certainly true that digital techniques are more readily applicable to these problems.

The disadvantage of sharing a common processor is unlikely to be a large consideration, provided failure is properly allowed for (e.g. power freeze or safety power levels) and provided the design and diagnostic aids facilitate quick fault analysis.

A major requirement on the temperature control unit is ability to accept data from the computer, as outlined in Section 2. An added bonus with a digital scheme is that these values do not have to be converted to analogue terms for error-forming purposes. The conversion of measured value signals into a digital form is, however, essential if these signals are to be monitored by the computer. Thus, given the right economics, the best technique appears to be a multi-loop digital package, taking as inputs: (i) the measured value in a digital form (analogue to digital (A–D) conversion possibly internal); (ii) setpoint values in a digital form; (iii) possibly additional inputs such as control terms, alarm setpoints and power limits; and giving as digital outputs: (i) alarm signals for excessive deviation from temperature; (ii) alarm signals for electronic failures; and also continuous analogue output signals to the output units controlling power to the heater bands. Obviously, digital inputs and outputs should be stored on the machine; this will be examined in Section 10.

The areas of temperature control will now be very briefly considered:

(i) *Barrel zones:* Of prime importance is the correct siting of the thermocouple relative to the heaters, i.e. mounting them as deep in the barrel as is practical. Where cooling systems are required (this is more common with extruders) the dynamics of these should be very carefully considered relative to the heating system so that proper three-term control can be applied both to heating and to cooling.

(ii) *Hot runners:* These are being used increasingly to improve the flow of plastics into difficult shaped mouldings; this applies particularly to multi-impression tools. Three-term settings will often require changing with different tools so digital techniques are useful here.

(iii) *Mould:* Temperature control of both halves of the tool is essential for high-quality, consistent mouldings. Thermocouples should be mounted in the tool for best results.

(iv) *Hopper:* Moulding consistency can be greatly improved by blowing temperature-controlled air through the material fed to the hopper. The temperature could be close to the plasticising temperature of the material.

(v) *Melt temperature:* Cascade control of melt temperature has been very successfully applied to extruders. The problems with applying this to injection moulding are those concerned with obtaining a reasonably priced reliable and accurate transducer. This topic will be dealt with more fully in Section 12.

3.2 Practical Implementation

Figure 5 shows a block diagram of the temperature control unit. The input conditioning and power output stages are covered in Sections 8 and 9 and

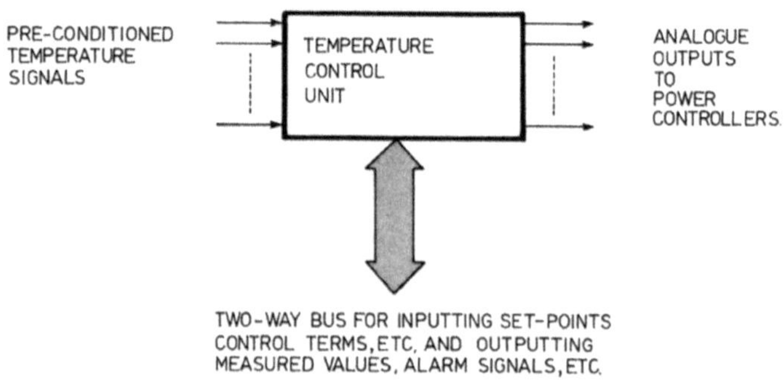

FIG. 5. Temperature control unit—block diagram.

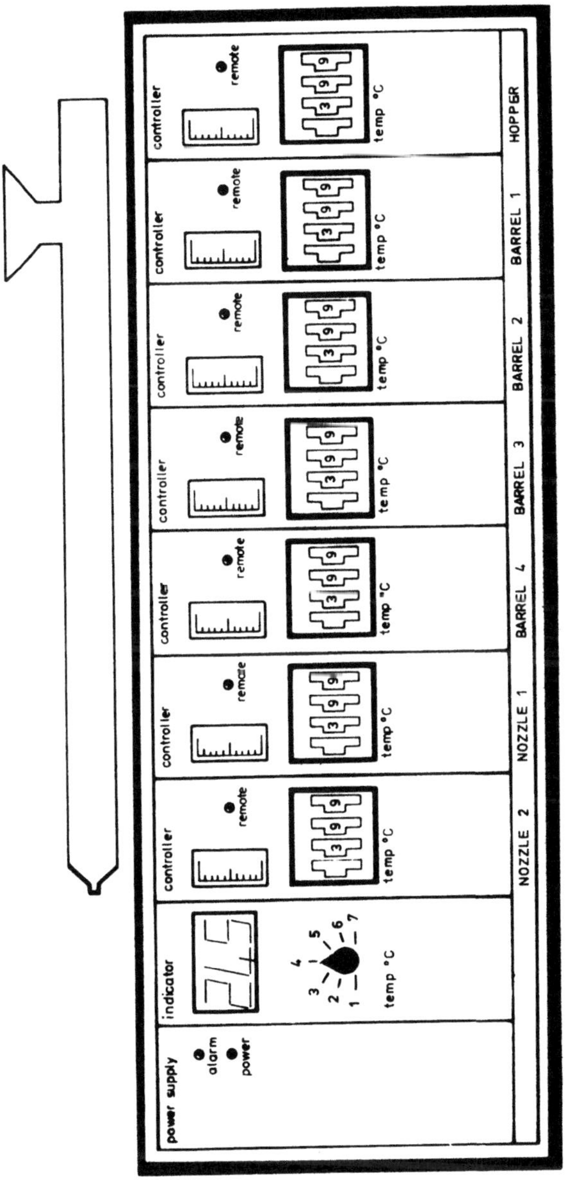

FIG. 6. Temperature control unit—operator controls.

the communications with other units on the machine and with the computer in Section 10.

A control panel using digitswitch settings is shown in Fig. 6 where the controls have been arranged ergonomically. Alternative methods of setting are discussed in Section 10.

4 VELOCITY/PRESSURE PROFILE CONTROL UNIT

4.1 Justification for this Unit

As stated in Section 2, it is argued that, to achieve best results, a velocity/pressure profile unit should be used for the injection unit. The justification for this statement will now be examined in more detail for although most people now accept the need for accurate control of barrel and nozzle temperatures, the necessity for accurately controlling pressures and velocities is still questioned. Therefore, before investigating methods of implementing such control, the question must be answered: Why control pressures and velocities in the injection unit?

In order to achieve consistently good mouldings, the requirement is that the speed of plastic entering the mould is such that a constant velocity wavefront is generated until the mould is full. This means the melt experiences a consistent freeze rate. This is best achieved by controlling the ram velocity with a profile which is determined by the characteristics of the plastic and the shape of the mould. Once the mould is full, pressure control is required during the setting and shrinkage of the product. The benefits obtained from controlling in this fashion relate to:

(i) consistency—a product with more repeatable and stable characteristics is produced;
(ii) strength—correct settings will ensure each product has reduced internal stresses and, hence, greater strength, less distortion and greater dimensional stability;
(iii) surface finish—by good initial injection;
(iv) material savings—overpacking of the mould can be avoided and less deviation in weight variations achieved;
(v) cycle times—overall mean cycle time can be reduced and deviations in this value made significantly smaller.

4.2 Practical Implementation

The exact implementation of the system will depend on the hydraulic system, but all systems have the following control strategy in common:

(i) During injection the actual velocity rate of the ram is under closed loop control by modulating the hydraulic oil flow to the injection ram by means of a servo valve. This eliminates the effects of material variation, namely basic material change (e.g. variation in percentage regrind), and changes in the machine (e.g. oil viscosity and ram friction). The hydraulic pressure will vary to achieve the desired velocity but should this exceed settable limits, it is normally arranged to change to pressure control.

(ii) As soon as the mould is full, either hydraulic or cavity pressure (discussed later) is controlled. To detect the condition of mould full, it is normally not sufficient to use ram position because viscosity variations of the melt and compressibility of molten plastic mean that a given ram movement will not always project the same shot size into the mould. A better, more reliable measurement is based on pressure and uses a pressure trip setting to actuate the change from velocity to 'hold-on' pressure control, using either hydraulic or cavity pressure (see later).

Figure 7 shows implementation of the system on a typical machine. This is not necessarily the best hydraulic arrangement (see Section 12) but is probably most typical; it requires minimum machine modification and, in many cases, is quite adequate.

Ram position is measured using a linear stroke transducer and the position signal is differentiated to produce a velocity signal. Long-life and noise-free signals can be obtained using a conductive plastic slide-wire (this is available commercially in lengths suitable for all sizes of machine). Velocity control is achieved via the servo valve (port A) with the relief valve closed.

As previously mentioned, the pressure trip may be worked either from the mould cavity pressure or from the hydraulic pressure. Mould cavity pressure has the advantage that it is a direct measurement of conditions in the mould and the pressure rises very steeply at the mould-filled point. It does, however, require considerable experience to position the sensor correctly within the mould. The normal measuring technique is to use a dummy measuring pin to transmit pressure to an external transducer. Every mould must then be fitted with this feature and it must be well maintained. Some moulds cannot be fitted with dummy ejector pins and for this reason (and because a machine-borne rather than a mould-related

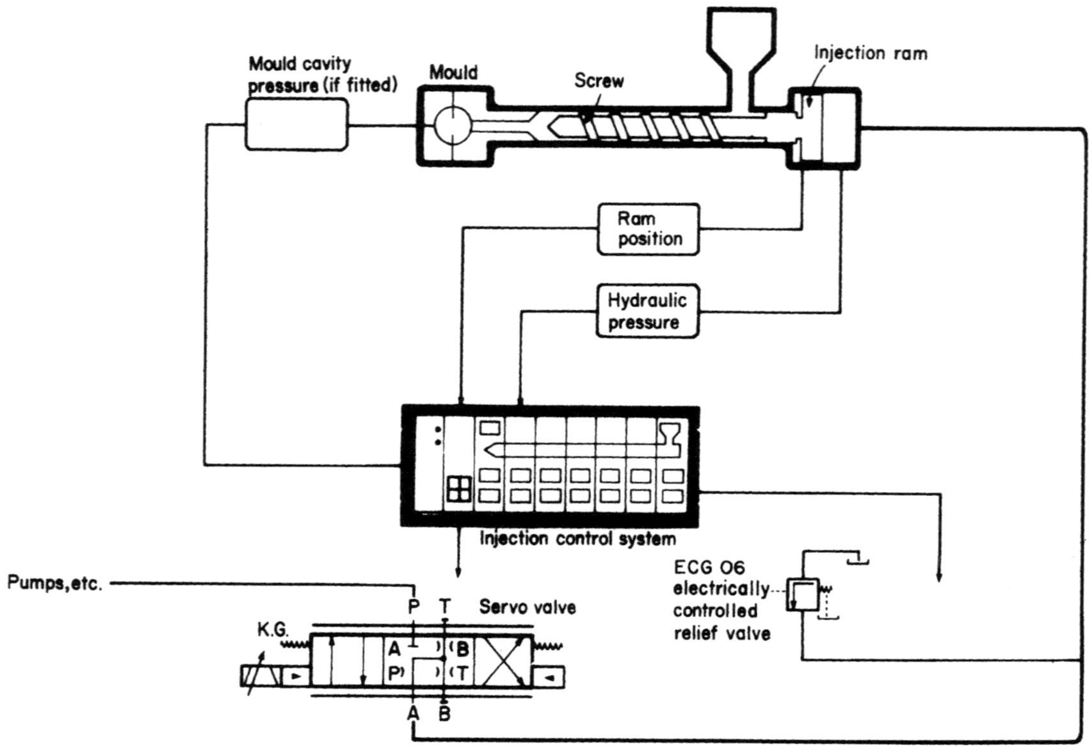

FIG. 7. Velocity/pressure profile control system—block diagram.

system is often preferred), a hydraulic pressure trip is often used instead. It has been found quite satisfactory and is much simpler and cheaper to trip from the hydraulic pressure.

Hold-on pressure is achieved using the electrically controlled relief valve and once this is complete, back-pressure is also controlled by this valve. On all counts, the hydraulic pressure measurement is best achieved by positioning the sensor just behind the ram. During screw-back, the screw speed can be controlled using the servo valve (port B) at either a pre-set or adjustable value. It is sometimes desirable to melt decompress at the end of screw-back and this is easily facilitated in the unit.

Obviously, more sophistications can be introduced, for example: (i) profiling hold-on pressure; (ii) profiling back pressure; (iii) profiling screw speed; and (iv) cushion control. The cushion is the possible ram travel left at the end of injection. Because the pressure trip is adapting the ram travel to compensate for melt variations, this means the cushion will also change. It is considered by some that it is important to keep the injection length constant and so adjust the ram backstop in order to keep the cushion constant. This means that the melt zone within the barrel is also being changed and this in itself will tend to change the melt properties. Other methods for obtaining constant volume are possible (Section 12). Cushion control will only be justified if the changes in cushion size are a significant percentage of the stroke, for example on small machines with screws having small length/diameter ratio (there is a tendency to make this ratio larger in modern machines).

Justification of any or all of the above may be possible on very critical applications but it is felt that the basic system is an extremely cost-effective one.

Using analogue techniques, two separate controllers are required, each having the ability to track the other's outputs in order to achieve bumpless transfers at the transitions of pressure/velocity control. Details of the servo valve and pressure relief valve drivers are given in Section 9. The logic signals which are necessary are: (i) signal to start injection—this can be used to reset the hold-on timer; (ii) signal to initiate screw-back (this is combined with the internally generated 'hold-on/timed out' signal). Pre-conditioning of input signals is covered in Section 8.

A control panel using digitswitch settings and selector switches for pressure trip and melt decompression is shown in Fig. 8. There are five velocity settings (with five corresponding position settings) and single set-points for trip pressure, hold-on pressure, hold-on time, back pressure and screw speed.

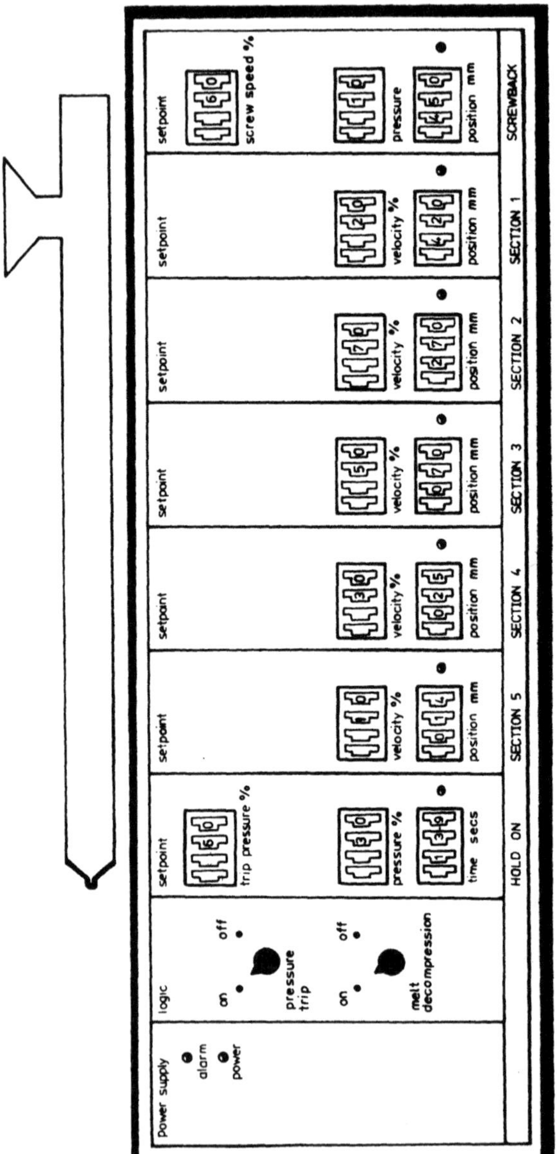

FIG. 8. Velocity/pressure profile unit—operator controls.

4.3 Setting up the Velocity Profile

The selection of injection speed is a compromise between speed and product quality. High injection speed means that the ram speed at the end of the injection stroke is so high that the mould can be overpacked. All this does is waste material, as an overweight, overfilled moulding does not benefit anyone, except the material manufacturer! This also tends to cause distortion and produces a moulding with high internal stress and relatively poor dimensional stability. Reducing the injection speed will improve the quality but it lengthens the cycle time and increases the chance of a short shot. Injection control allows a speed profile to be set up so that a high injection speed can be used for most of the injection but this can be reduced to a slow speed over the final portion of ram movement, in order to prevent overpacking of the mould, as well as allowing a smooth transition into the hold-on phase. Hence, short cycle times can be achieved without overpacking the mould. Mould flash is also significantly reduced.

Jetting, a problem which takes place in some mouldings, particularly those with narrow cross-sections, is where the material squirts, rather than flows, into the cavity at the beginning of injection because of the absence of any back pressure within the mould cavity. This can be countered by setting a slow speed over the initial part of the injection stroke in order to establish a smooth flow of material into the mould. The velocity profile can be adjusted to suit the particular mould. Simple moulds do not normally need more than three speed sections in the profile, but more complex mouldings can benefit from having up to five sections. This allows the injection speed to be varied so that the speed of the melt front in the mould can remain near constant, even if the cross-sectional area of the mould changes significantly. The smooth, even flow of material into the mould enhances the surface finish of the moulding as well as reducing the internal stress, thereby increasing strength and improving dimensional stability.

5 PROGRAMMABLE LOGIC UNIT

5.1 Basic Principle

As mentioned previously, the fitting of a velocity/pressure profile unit to the mould unit is usually more difficult to justify and would depend on the design of the mould unit. The more complex the functions performed by the mould the more likely the justification.

Let us assume, therefore, that the mould is fitted with a programmable logic unit. There are certain logic features common to the mould and injection unit and it is sensible, although not essential, to combine these also. (In fact, even with profiling units on both sections, a simpler form of logic unit would be necessary.)

As with temperature control, the techniques of solid state logic systems are covered in detail in Chapter 9. Therefore, only those additional features necessary in order to achieve the objectives set out in Section 2 will be discussed.

The requirements of this unit are that it must be able:

(a) to accept digital inputs from limit switches and pressure switches on the mould (Section 8);
(b) to accept time set-points from an operator panel (discussed later) or from the central computer, such as screw delay, cooling, recycle pause, ejector delay, etc.
(c) to accept digital inputs from mode selector switches on the operator panel (see later);
(d) to accept digital inputs from the manual control devices on the operator panel;
(e) to perform the necessary sequence, based on the previous signals input and to give out the necessary signals to the power control devices on the mould (Section 9). Logic signals must also be put on to the internal, parallel bus (for communicating to the other units on the machine monitoring and control unit) and on to the external serial bus (for communicating with the mini-computer).

The amount of information to be communicated to the mini-computer will vary depending on its function (Section 11). However, for comprehensive monitoring, the entire memory and the input–output (I/O) contents of the programmable logic unit should be made available to the computer via the data link. This will certainly facilitate central fault diagnosis as well as comprehensive production data.

Entry of the sequence into the programmable logic unit is slightly more complex. It is very desirable that this can be done locally, either by a resident panel which can be locked away during normal operation or by a hand-held device which is plug-in. These devices could have access on to the parallel bus, the serial bus or, possibly, on to some internal bus structure. If it is not required to change the sequence from the central computer, the choice would not matter too much.

Entry of sequence program from the mini-computer is only likely to be worth while when there are a large number of machines and a large number of similar moulds. If entry via the main bus is not required, it is probably better to transfer the sequence on to more permanent memory than that used for the overall unit (Section 10). Two possibilities are Programmable Read Only Memory (PROM) and Electrical Read Only Memory (EROM), with plug-in blocks which can be changed and stored away for different moulds.

It is not intended to describe the mould sequence in detail. Normally the mould close sequence will be split into slow–fast–slow and the mould open likewise. Other functions to be covered are ejector forward and retract and core operations. For the injection unit there are functions such as carriage forward and retract and nozzle open and close. There is also pump on/off selection and functions connected with the gate.

5.2 Practical Implementation

A block diagram of the unit is shown in Fig. 9 and the main features of the sequencer are listed below:

(i) Memory—size will vary widely for different machines but is typically 256 words. Memory type is most likely to be a mixture of read–write and PROM.
(ii) Logic—usually functions involving shift registers, latches, etc.
(iii) Programmable up–down counters and timers—to handle auxiliary time functions in the mould cycle and to provide information on the number of machine cycles, etc.
(iv) Battery back-up—to retain read–write memory in the event of power failure.

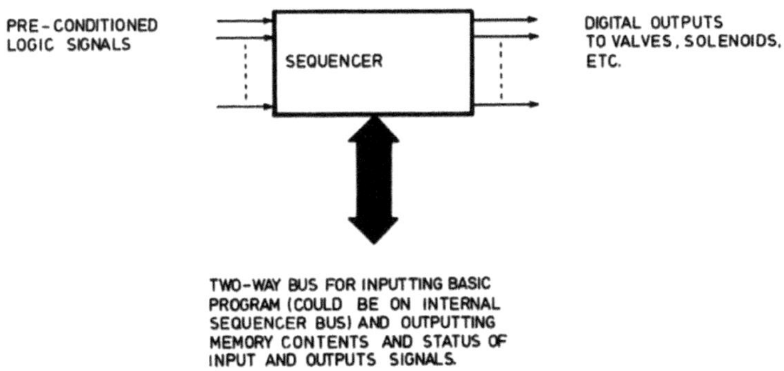

FIG. 9. Programmable logic controller—block diagram.

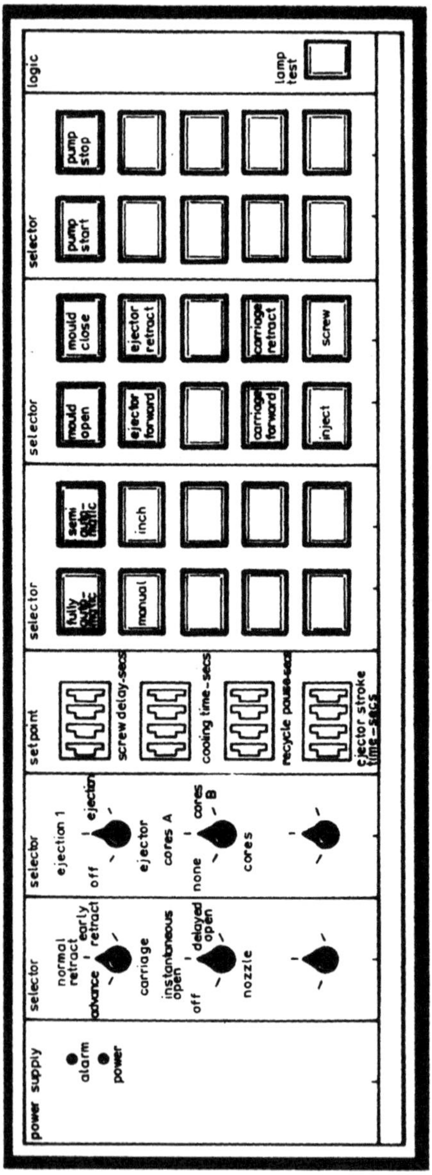

FIG. 10. Logic—operator panel.

Whatever device is used for programming (see p. 247), entry will be via a keyboard. The method of programming is usually by ladder diagrams (relay or Boolean terminology) or flow diagrams. Ladder diagrams have restrictions (not serious for simple machines) but are more easily understood by operators used to relay logic.

A control panel using digital switch time settings, mode selector switches and a pushbutton panel for manual operation is shown in Fig. 10. Alternative presentations are discussed in Section 10.

6 ALARM ANNUNCIATOR UNIT

6.1 General Problem

The situation regarding alarms on an injection moulding machine is a complex one. As stated previously, it is also a very important one because of the damage that could result if due care was not taken in this area. Since the alarm strategies are complex and will vary from machine to machine and to some degree with tool changes, etc., a microprocessor-based unit performing in a consequential rather than sequential (Section 5) manner is ideally suited to the task. The alarms can be conveniently categorised:

(i) *Priority alarms*—those which must be actioned immediately. Examples of this type of alarm are: (a) motor overload tripped, (b) mould obstructed, and (c) injecting into full mould.

(ii) *Semi-priority alarms*—those which can be actioned when present injection cycle is completed. Examples are: (a) faults in the oil circuit such as low level, overtemperature, suction filter blocked, return filter blocked, etc., (b) mould overtemperature, and (c) injection unit overtemperature.

Mains failure needs special treatment. Providing non-volatile stores are used in the various units, the machine could start up automatically following a short mains failure. Otherwise, information as to where the machine was in the cycle before failure would be lost and operator intervention would be required. Long mains failure would always necessitate the intervention of an operator.

6.2 Detection of Alarms

The first source of alarm information is the units described in Sections 3, 4 and 5. These alarms will be of two types, unit failure alarms and process alarms. With respect to the first of these, as most of the units have been designed with a large degree of intelligence, it is obviously possible, by careful design, to build in a fair degree of self-diagnostics which will allow, say, the alarm annunciator not only to show which unit is faulty but also what type of fault has occurred. With regard to process alarms, some examples are given below:

Temperature control unit. Detection of excessive deviation of temperature is an extremely useful feature. Temperature too low could be used to inhibit any screw movement in order to avoid damage to the screw with cold plastic in the barrel. Temperature too high would obviously result in burning; this is an example of a case in which diagnosis is necessary since this condition has several causes, e.g. thyristor lock-on in power device, excessive work-heat in the barrel giving excessive temperature even with no heater current, faulty three-term action, etc. Given the appropriate information, the cause of overheating can be diagnosed by the alarm annunciator unit. Absolute temperature alarms may also be felt necessary, e.g. key temperatures such as tool temperature (fixed and moving halves) and melt temperature may be selected.

Velocity/pressure profile unit. Excessive deviation on velocity and pressure controllers will show the profile is not being achieved (obviously these have to be examined at the appropriate part of the cycle). It may be permissible, however, to allow transient alarms such as may occur on changing from one velocity section to the next, hence requiring timing of these alarms before taking any action. Again, absolute alarms, say, on cavity pressure, may be provided, for example, to stop injection into full mould, and checking that the first stage of injection does not exceed a pre-set limit.

Programmable logic unit. Supervision of this unit is required in some degree in order to detect excessive deviations from the desired sequence.

The second source of alarm information is from sensors fitted for this purpose. Types of fault have been covered in Section 6.1 and examples of sensors are: (i) thermal trip on main motor; (ii) micro-switch for mould fully closed plus overall mould close timer; (iii) fully retracted limit switch for injection unit; and (iv) level, temperature and pressure switches in the oil circuit.

6.3 Practical Implementation

There are many ways in which an alarm unit can be achieved. However, with regard to the injection moulding machine, it is the preference of the author that all analogue variables, where absolute alarms are considered essential, should be detected within the input conditioner (see Section 8). This can also cover transducer failures, for example, an open-circuit thermocouple. This means that the alarm annunciator accepts purely logic inputs. Its implementation and method of programming become very similar to the programmable logic unit in that it requires: (i) memory—to store transient alarm states and allow acknowledgeable and non-acknowledgeable states; (ii) logic—for the alarm strategy; (iii) counters and timers—for timing length of alarm state and counting to eliminate spurious alarms.

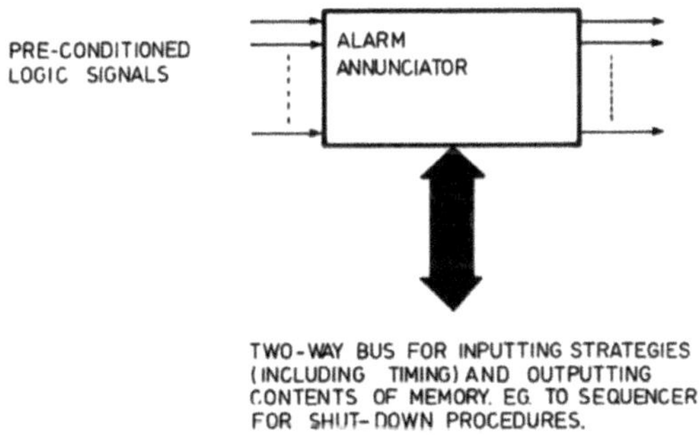

FIG. 11. Alarm annunciator—block diagram.

Supervisory timers for, for example, fill, hold and cycle times are also required; and (iv) battery back-up—to retain states and alarm for mains failure. A block diagram is shown in Fig. 11 and an operator panel (greatly simplified) in Fig. 12.

7 PRODUCTION MONITORING UNIT

The function of the production monitoring unit is to take information from the units previously described plus information that must be entered by the operator (automatic weighing, surface finish inspection will not be considered at this stage—see Section 12) and perform simple arithmetic tasks such as allowing basic production data to be displayed locally. More complex data involving statistics and economics would be performed centrally (see Section 11).

Information that must be entered manually includes such terms as: (i)

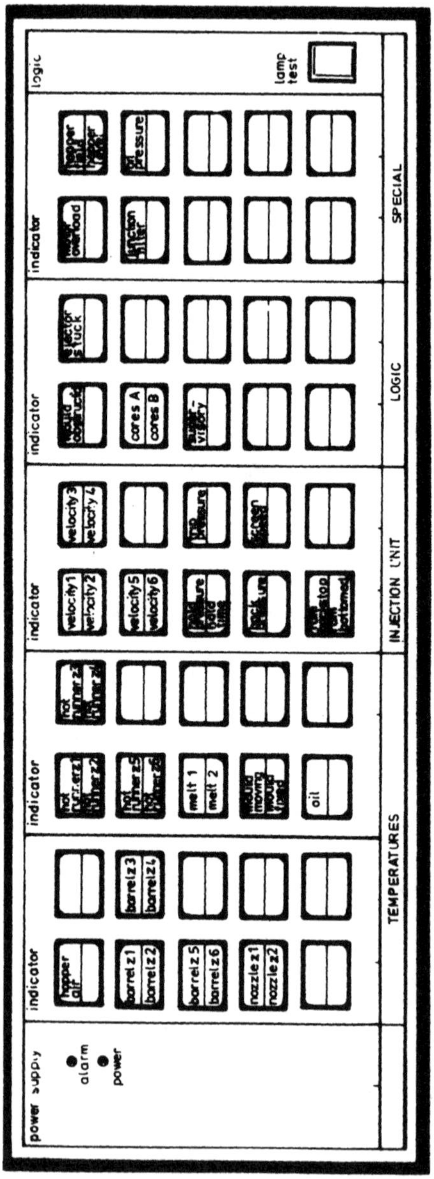

FIG. 12. Alarm annunciator—operator panel.

machine number; (ii) tool number and number of cavities; (iii) job title; (iv) shot weight; (v) quantity of mouldings required; (vi) number of good mouldings per shot—entered by operator; and (vii) reason for machine stoppage where appropriate (this would in many cases be correlatable with the information from the alarm annunciator)—entered by operator. Items (i) to (v) would be entered by the supervisor and with a central mini-computer may always be entered centrally.

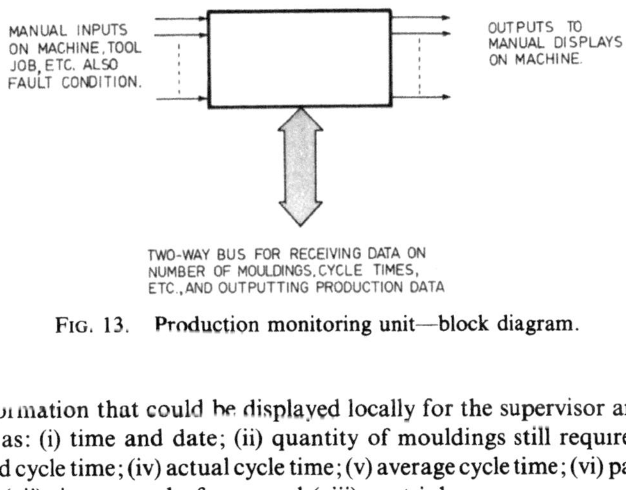

FIG. 13. Production monitoring unit—block diagram.

Information that could be displayed locally for the supervisor are such items as: (i) time and date; (ii) quantity of mouldings still required; (iii) desired cycle time; (iv) actual cycle time; (v) average cycle time; (vi) parts per hour; (vii) time to end of run; and (viii) next job.

Figure 13 shows the block diagram and Fig. 14 the operator controls.

8 INPUT CONDITIONING UNIT

8.1 Basic Units

The principal input signals required for the units described in the previous five sections are shown in Tables 1A and B. They are categorised firstly into analogue (Table 1A) or digital (Table 1B) signals. Analogue signals are further divided into cyclic and non-cyclic variables; digital signals are subdivided into status or alarm signals. Finally, the location on the machine (injection or mould unit), and the type of transducer required are given.

The basic units required for pre-conditioning the analogue and digital signals will now be described.

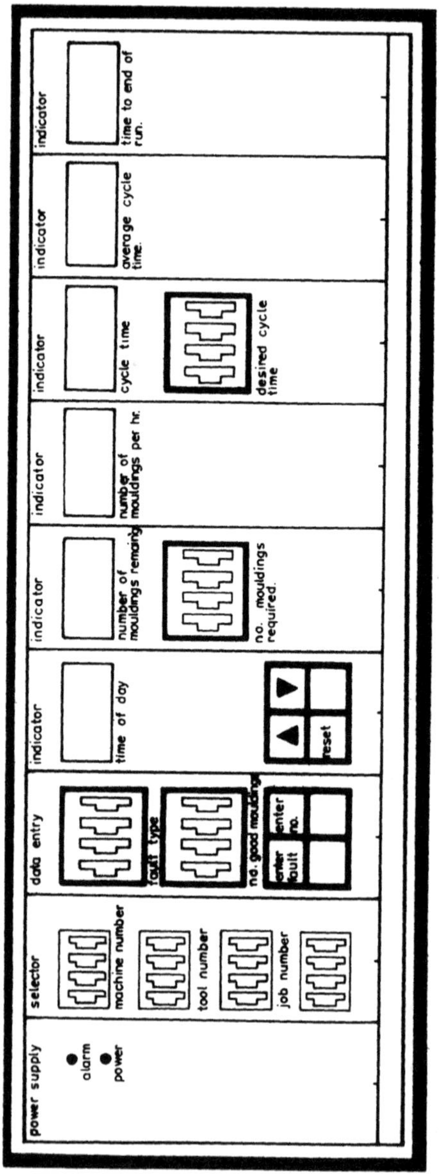

FIG. 14. Production monitoring unit—operator controls.

TABLE 1A

BASIC INPUT SIGNALS (ANALOGUE)

Analogue inputs	*Cyclic or non-cyclic*	*Injection unit or mould unit*	*Most common transducer*	*Comments*
Barrel zone temperatures	Non-cyclic	Injection	Thermocouples	Usually between 4 and 12 zones
Melt temperature	Cyclic	Injection	Thermocouple	Requires special device (see text)
Hydraulic pressures	Cyclic	Injection	Strain gauges	Injection pressure/back pressure
Melt pressure	Cyclic	Injection	Strain gauge	Requires special device (see text)
Ram displacement	Cyclic	Injection	Plastic slidewire	Derive velocity signal from this
Screw speed	Cyclic	Injection	Tacho	Not usually closed-loop controller
Screw temperature	Non-cyclic	Injection	Thermocouple	Not common on injection machines
Hopper air temperature	Non-cyclic	Injection	Thermocouple	Control close to plastication limit
Hopper level	Cyclic	Injection	Capacitance	Could be digital (alarm) only
Hot runner temperatures	Non-cyclic	Mould	Thermocouples	Used with difficult shapes
Hydraulic pressure	Cyclic	Mould	Strain gauge	Do not usually continuously monitor
Cavity pressure	Cyclic	Mould	Strain gauge	Special device (see text)
Tool temperature	Non-cyclic	Mould	Thermocouple	Separate devices in fixed and moving halves

TABLE 1B
BASIC INPUT SIGNALS (DIGITAL)

Digital inputs	*Status or alarm*	*Injection unit or mould unit*	*Most common transducer*	*Comments*
Back-stop on ram	Alarm	Injection	Limit switch	To detect overshoot
Carriage forward	Status	Injection	Limit switch	
Carriage back	Status	Injection	Limit switch	
Nozzle valve	Status	Injection	Micro switch	
Mould forward movement	Status	Mould	Proximity switches	Usually three for slow–fast–slow
Mould backwards movement	Status	Mould	Proximity switches	Usually three for slow–fast–slow
Overstroke	Alarm	Mould	Limit switch	To detect overshoot
Ejector position	Status	Mould	Limit switches	
Cores	Status	Mould	Limit switches	
Gate	—	—	—	

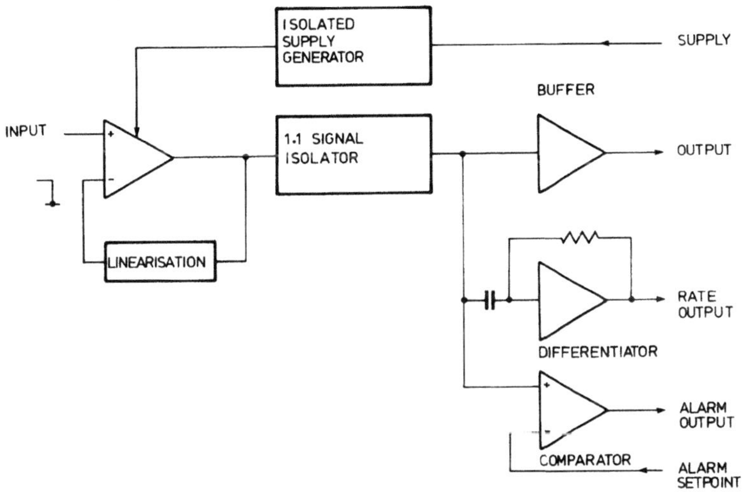

FIG. 15. Input convertor—block diagram.

Input Converter (for Analogue Signals)

The block diagram for the converter is shown in Fig. 15. The main features of the unit are:

(a) wide range of inputs, i.e. to cover all variables contained in Tables 1A and B;
(b) adequate accuracy and stability—for the injection machine accuracies of $\pm 0{\cdot}5\%$, stabilities $\pm 0{\cdot}25\%$;
(c) isolation and linearisation of the input—isolation of up to 240 V a.c. for all signals is essential for non-isolated transducers and very desirable even in cases where the transducer is mechanically isolated (e.g. strain gauge, position sensor) for safety reasons;
(d) output of 0–10 V— this is a convenient level to normalise to, being high enough for noise immunity and within the range of modern semi-conductors;
(e) rate measurement circuit—an optional feature used for velocity measurements; also possible to detect excessive rates of rise of, say, pressure;
(f) alarm circuit—allowing full-scale or deviation alarms for critical parameters. Alarm set-points can be pre-set or fed in remotely if

these set-points are likely to change for different moulding conditions. Alarm light-emitting diodes (LEDs) on the units aid fault-finding.

Opto-isolator (for Digital Signals)
Optical isolation has become the standard method for treating logic signals. Such devices provide a means for electrically isolating two signals using, say, LEDs. The block diagram for the isolator is shown in Fig. 16. The main features of the unit are:

(a) sixteen channels of opto-isolation—each input is buffered with pull-up or pull-down resistors chosen with regard to fail-safe operation;
(b) high voltage possible between channels—again, mains voltages must be allowed for;
(c) input filtering for noise immunity and 'anti-bounce'—these two features are essential with injection moulding machines as failure in either respect would have serious consequences;
(d) several input and output voltages available—compromises between safety and reliability and differences in standards make this feature essential;
(f) LED indication on input state—aids fault diagnosis;
(g) tri-state outputs—output can be gated on to a parallel bus (see later).

8.2 Analogue Bins
Packaging of the input converters can be in many forms—one possibility is shown in Fig. 17. This bin accommodates eight units plus interface units. Analogue outputs are scanned by a field-effect transistor (f.e.t.) scanner and the output is fed into a A–D converter. This has a 16-bit parallel output which is fed via an interface unit on to the bus. This unit also takes digital signals off the bus. These are fed via an A–D converter and demultiplexer unit to provide alarm set-points. The bus also has to handle data for addressing the unit (decoded locally) and ranging information, since it is very desirable (see later) that digital data is in engineering values. Satellite bins with 16 converters can be added.

8.3 Digital Bins
A similar concept to that shown in Fig. 15 applied to the opto-isolators gives digital bins of the form shown in Fig. 18. Eight isolators are fed on to a common 16-way bus which communicates via an interface unit on to the bus. Again, address information is locally decoded.

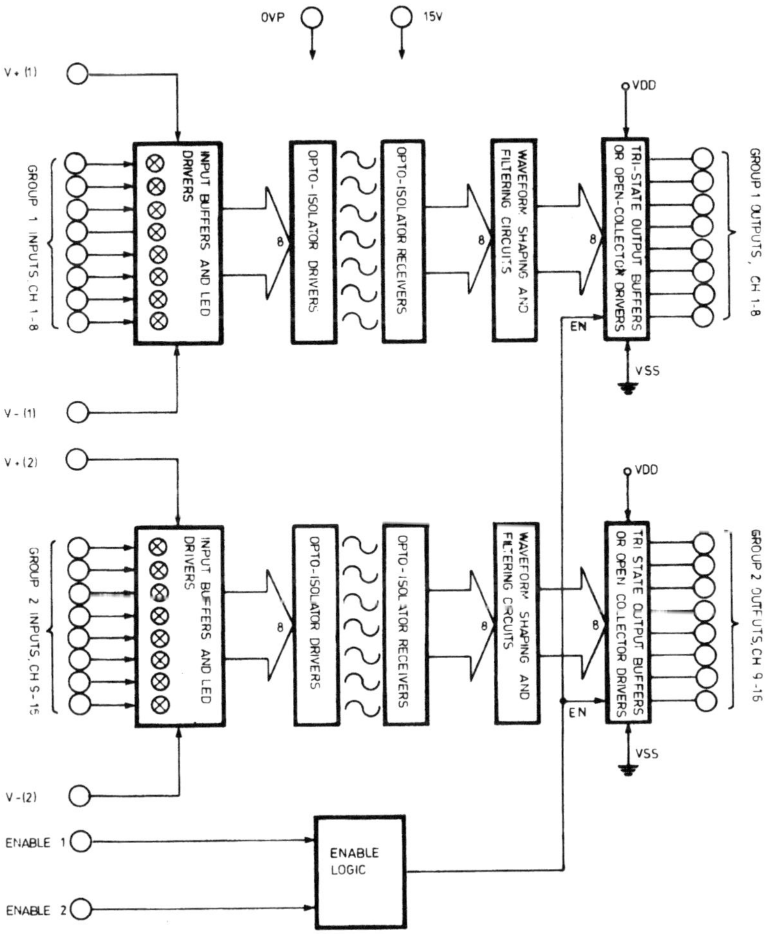

FIG. 16. Opto-isolator—block diagram.

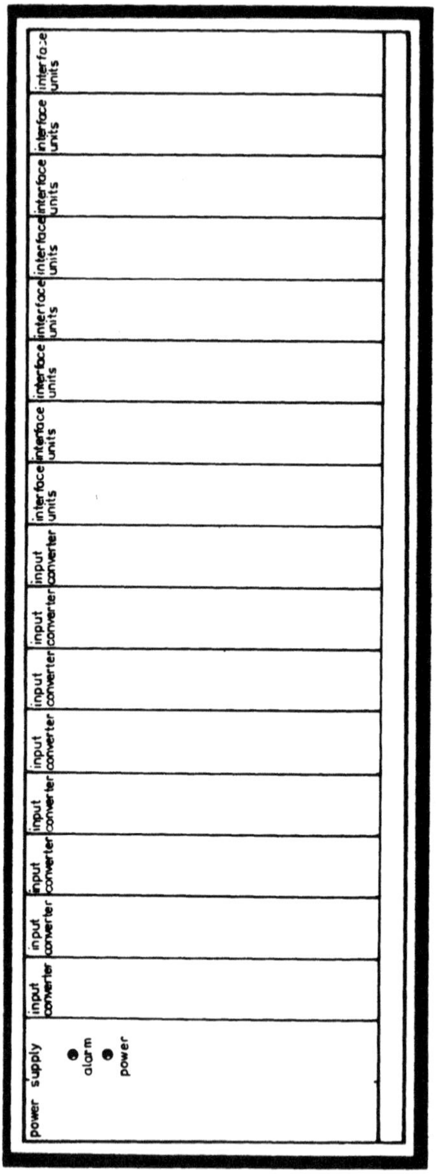

FIG. 17. Analogue bins.

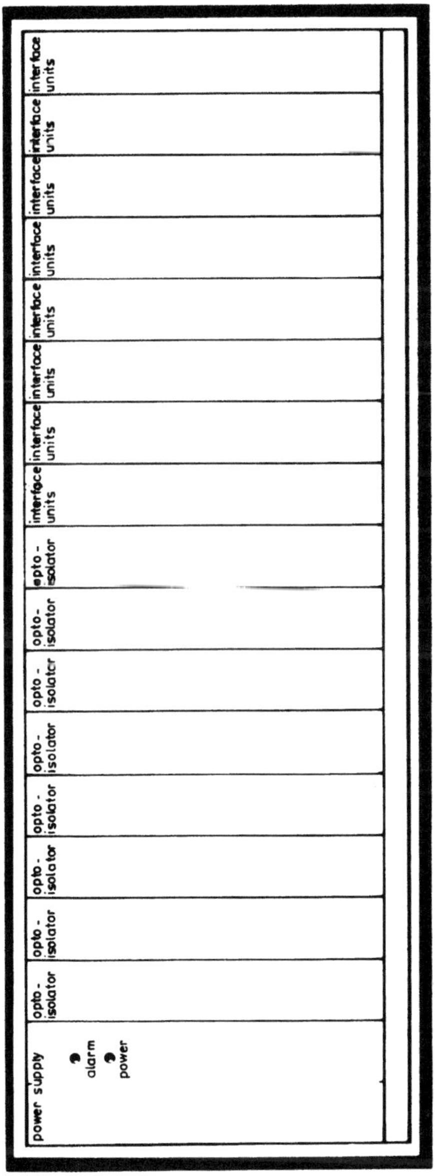

FIG. 18. Digital bins.

9 POWER CONTROL UNITS

9.1 Basic Unit

The four basic units for the control of electrical heating and hydraulic power are described below. Special features required by the overall philosophy outlined above are listed in each case:

(i) *Thyristor power controller*—for the barrel heaters and hot runners (see Fig. 19); main features:

(a) up to 75 A, 415 V—this would cope with even the largest machines;
(b) fast cycle firing—phase angle would not be used for normal heaters;
(c) compensation for supply voltage variations—prevents disturbances on the temperature loops;
(d) self-powered—if the temperature control section fails, the unit can be arranged to fall to a safe limit (power freeze is possible but probably not worth the expense);
(e) fuse fail indication and logic output to feed the alarm annunciator;
(f) optional logic signal for power demand—allows load failure detection in the alarm annunciator;
(g) optional 0–10 V outputs for load voltage and current monitoring (load fail or continuous monitoring);
(h) optional cool channel with triac output for operating water-cooling solenoids or air-cooling fans in slow cycle switching mode—these would incorporate compensation for the non-linear nature of these devices (not shown on diagram).

(ii) *On–Off controller*—for contactors; main features:

(a) up to 1 A, 240 V output, fully isolated;
(b) time-proportioning or on–off modes;
(c) zero voltage switching—to minimise interference generation;
(d) compensation when used for cooling (see above).

(iii) *Proportional driver*—for servo and pressure relief valves (see Fig. 20); main features:

(a) ± 50 mA bi-directional fully isolated output with appropriate response times;
(b) settable limits for maximum and minimum swings;
(c) adequate linearity between input and output.

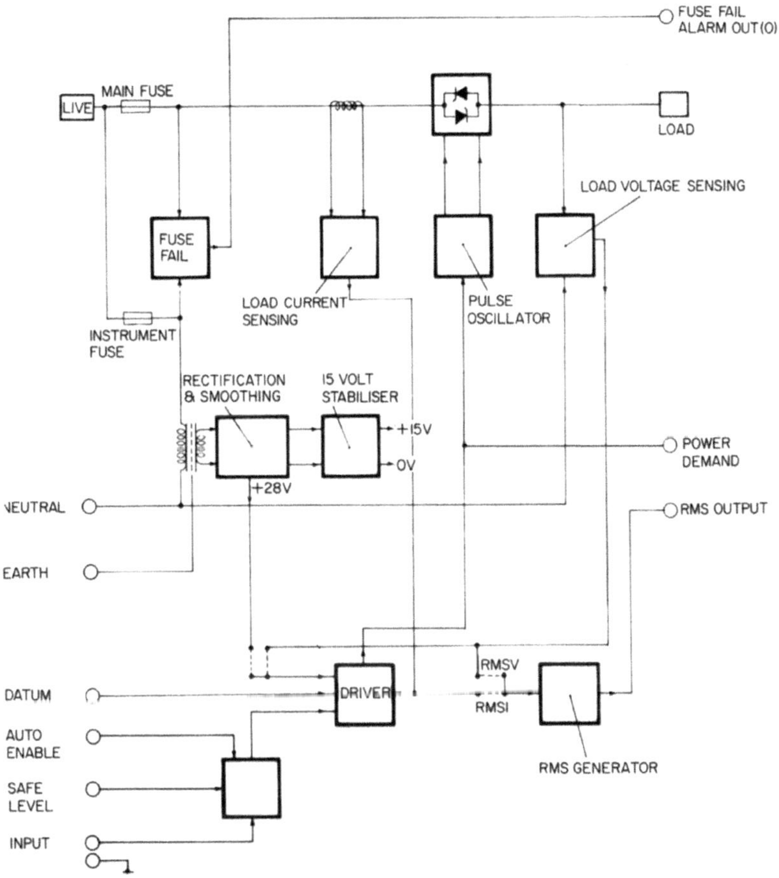

FIG. 19. Thyristor power controller.

(iv) *On–Off driver*—for on–off hydraulic valves; main features:

(a) fully isolated d.c. output rating up to 48 V, 0·5 A;
(b) able to cope with inductive loads;
(c) appropriate speed of response.

For all the above units, protection of the output is essential. This can be done simply by appropriate fusing but more elegant circuits are possible

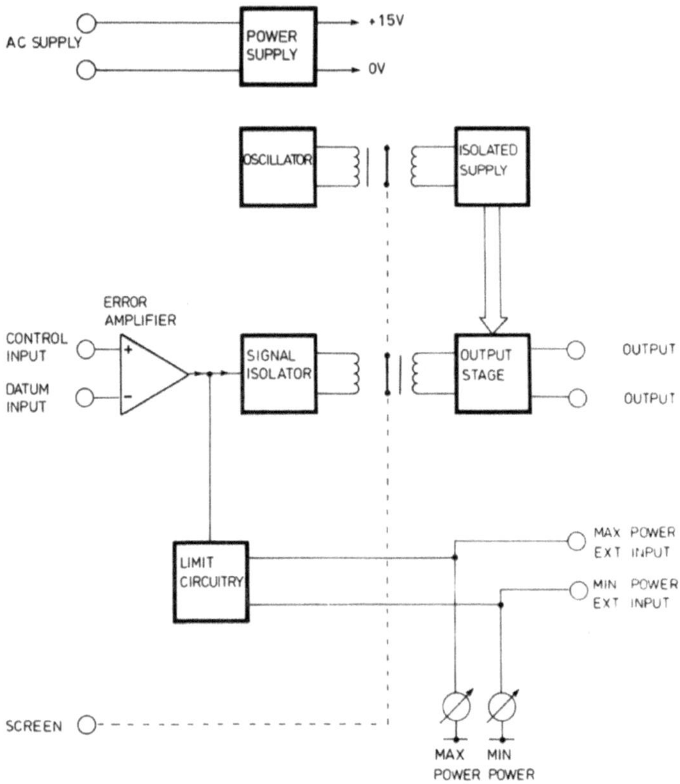

FIG. 20. Proportional driver for servo valves.

which are usually of a 'trip-and-try-again' nature, i.e. the output is disabled immediately a short circuit is established and then will attempt to re-establish itself periodically until the short circuit is removed.

10 MACHINE MONITOR AND CONTROL UNIT

10.1 Basic Unit

The basic scheme for the complete machine monitor and control unit is shown in Fig. 21. The five basic units described in Sections 3, 4, 5, 6, 7, i.e. temperature controller, velocity/pressure profile controller, programmable

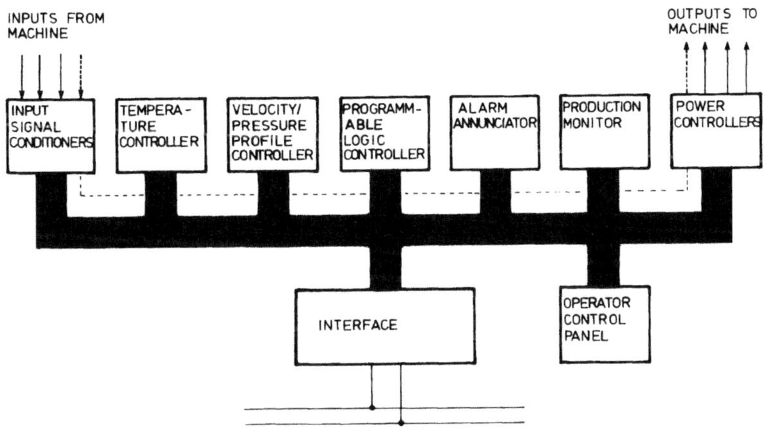

FIG. 21. Machine monitor and control unit—block diagram.

logic controller, alarm annunciator, and production monitor, together with the input signal conditioners and the power controllers, all connect on to a parallel bus which does not extend beyond a single machine. This bus is taken to an interface unit which connects on to the serial bus which communicates to the central mini-computer.

The optimum form for this interface is that it contains non-volatile memory containing:

(i) all analogue set-points which are designed to be fed from the central mini-computer (this may not be done on initial installations—see later);
(ii) all analogue measured values required by the central mini-computer for production monitoring and control (see Section 11);
(iii) all digital status words for determining machine logic that is likely to be sent from the central mini-computer;
(iv) all digital status and alarm information required for production control.

This intermediate store and associated control is designed so that the computer has immediate access to any data, never having to wait for conversions, etc. The main point then to be determined is how to update the data in the store (this is not really a problem with the relatively slow injection process). Another feature obtained using storage in the interface is that data are retained even with electronic failure of any unit.

A further, very important feature is the simplification of operator control that can be achieved. It is obviously essential that set-point data can be entered locally, primarily because the unit has to work without the mini-computer for two reasons: (i) installation of the unit may be ahead of the installation of the mini-computer; and (ii) failure of the mini-computer must not stop production. Obviously, the digitswitch/selector switch type of panel already described could be retained, but the store allows a much simpler keyboard-type panel. This could well be of the conventional teletype pattern and it is not ridiculous to consider visual display units (VDUs) on each machine. However, it is the author's view that only one portable VDU terminal is necessary for initial setting-up (i.e. entering sequence programs, changing controller three terms, etc.) and that the panel on the machine should be more readily understood by the average machine operator and supervisor. A typical panel embodying all the entry and display functions for the five basic units will now be considered.

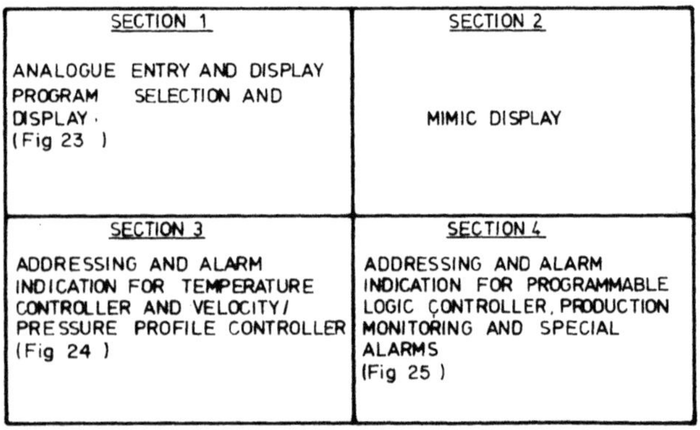

FIG. 22. **Machine monitor and control unit—control panel.**

10.2 Local Display/Entry Panel

Figure 22 shows a possible scheme for a local display/entry panel. The panel is divided into four main areas which will be discussed in turn:

(i) *Analogue Entry and Display, Program Selection and Display*

Figure 23 shows a typical panel. Considering first analogue variables, once a loop is addressed (see below) there is a digital display of: (a) loop number, (b) set-point (for a control loop; for a monitoring loop this could be blank or show a desired or alarm value), (c) measured value, (d) units (°C, mm,

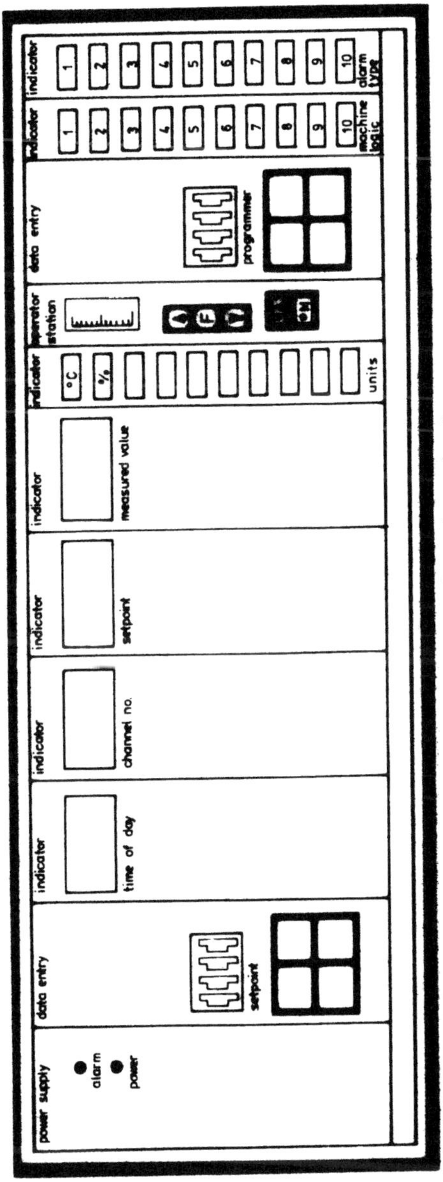

FIG. 23. Control panel—section 1.

etc.), and (e) power level, where relevant (heater current, valve current, etc.). Entry of a new set-point is achieved simply by dialling the new value (in engineering units) and pressing the entry button. A key and double pushbutton entry prevents unlawful entry. For the digital variables, selection of a function (e.g. melt decompress) would allow one of ten lights to be illuminated (for melt decompress, say, on for a short time, off for a long time). Interpretation of these status lights may need a table but this is not considered a disadvantage because of the infrequency of requiring this information. Principal alarms are shown on the addressing panel and there is a unit in this rack to show the type of alarm.

(ii) *Mimic*

It is not intended to describe a mimic diagram here as this is very individual. The LED is very useful for the mimic—red LEDs for alarms (flashing unacknowledged, steady acknowledged), green LEDs for status and flashing green LEDs to show analogue values (heater current, etc.). The mimic could also contain continuous analogue displays of key parameters.

(iii) *Addressing for Temperature Loops and Velocity/Pressure Profiling*

Figure 24 shows the addressing panel for temperatures and velocity/pressure profiles. All parameters covered in Sections 3 and 4 are inscribed on the pushbuttons which have split screens so that the top half is illuminated white (to show which loop is addressed) and the bottom half flashes red if that parameter goes into alarm (this goes to steady red on acknowledgement, thence going out once the alarm has gone).

(iv) *Addressing of Sequencer, Production Monitoring and Special Alarms*

Figure 25 shows the addressing panel for the sequencing/production monitoring and special alarm functions covered in Sections 5, 6 and 7. Entry of machine sequence is not possible by this panel—in line with the approach taken—but could easily be accommodated.

Finally, there is still the operator's manual panel (normally an operator would not need to use the panel described above), which also includes the digitswitch for fault type (see Section 7).

A system similar to the one described is shown in Fig. 26. This is on a machine at Rolinx Limited, part of the ICI Group. The system is installed on a machine for producing what is believed to be the world's largest moulding—the hull and deck of a 3·4-m sailing dinghy.

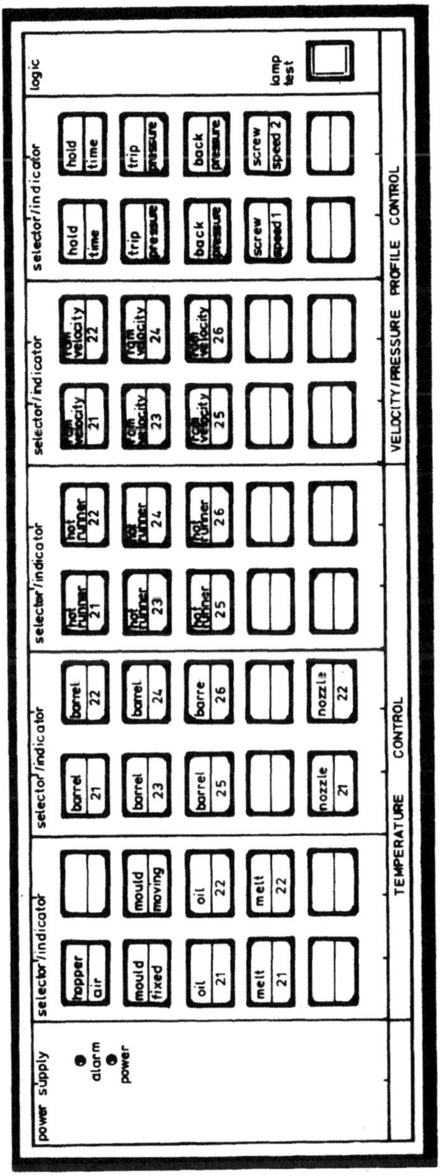

FIG. 24. Control panel—section 3.

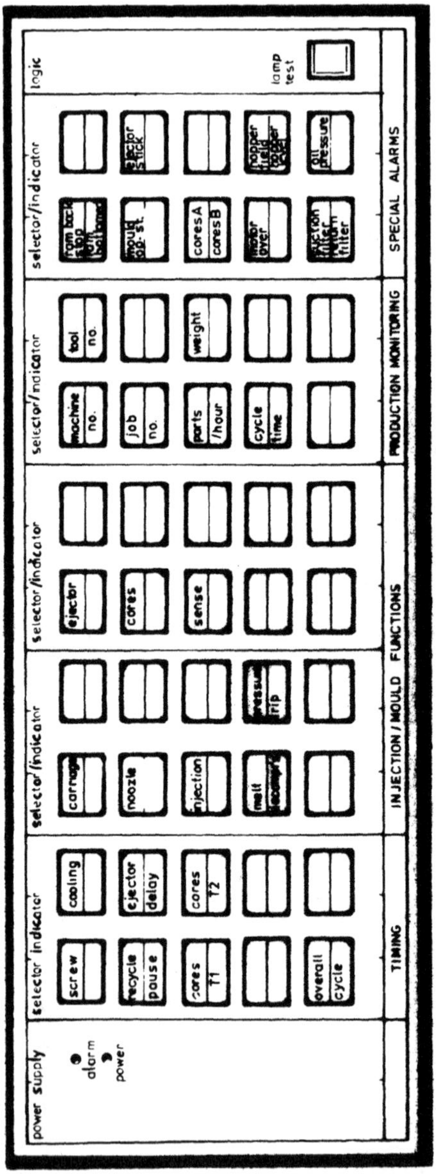

FIG. 25. Control panel—section 4.

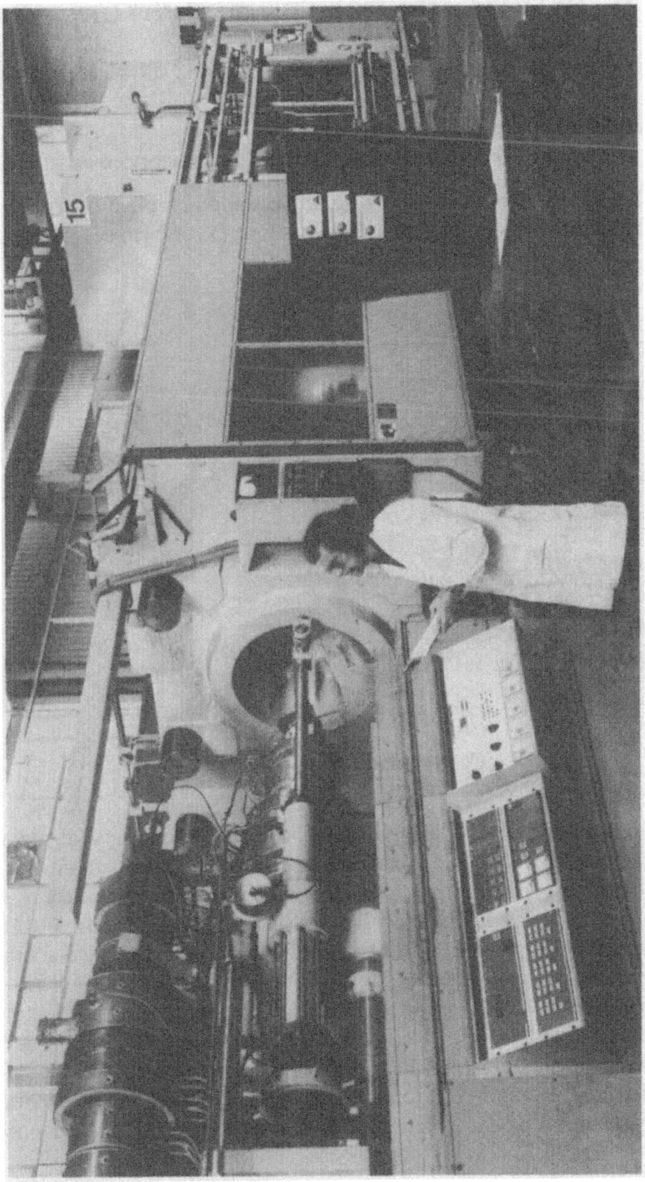

FIG. 26.

11 OVERALL MONITORING AND CONTROL

A diagram of a possible mini-computer installation is shown in Fig. 27.

11.1 Functions of Central Mini-computer

The functions of the central mini-computer can be categorised as follows:

(i) *Report generation and monitoring.* Information is collected from the individual machines for management reports. The exact form of this report

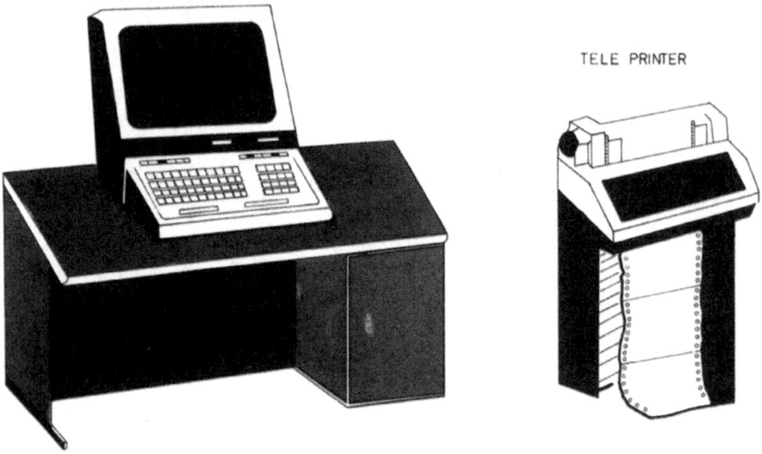

FIG. 27. Central mini-computer installation.

is considered later in this section, but basically it consists of those parameters covered in Section 7 with the addition of costing information plus key operating parameters from the individual units discussed in Sections 3, 4, 5, 6, 8 and 9. The reports will usually be displayed on a VDU and also periodically printed. It will be possible to interrogate the machine at any time to see its present status. Also, should any alarm condition arise, this can give audible warning plus automatic display on the VDU. Printing can also be initiated by the presence of an alarm and the subsequent disappearance of this alarm. The alarm strategy in this respect will be programmed into the mini-computer. Generation of alarms on the machine has already been discussed; additional alarm conditions can, of course, be performed centrally.

(ii) *Storage of operating conditions.* Operating conditions for a job previously done will be stored, say on magnetic disc, so that at the start of a new production run the machine number, tool number and job title are entered via the mini-computer keyboard (as discussed previously, this can be done on the machine). All operating conditions—pressures, velocities, temperatures, etc.—are transferred from the central to the local store on the machine (Section 10) and after the appropriate warm-up time, the machine will be automatically set up. With a completely new production run, the machine would most probably be optimised locally and once satisfactory results were obtained, operating data would be transferred from the local to the central store. It is also feasible to adjust control parameters automatically according to the information gathered (see above); this is covered later in this section.

11.2 Report Generation and Monitoring

The format of a report will, of course, vary markedly from plant to plant but would typically contain the following information for each machine:

(i) Machine status—time and date, machine number, tool number, number of cavities, job title, present status of machine. If the machine is stopped, the display would show the reason, this information being automatically detected or entered by the operator (Section 7).

(ii) Basic production data—quantity required, quantity produced, quantity remaining, time since start of run, time still to run, amount of down time and analysis of reasons for down time, details of next job and estimated start time.

(iii) Statistical data—comparison of actual cycle time against standard giving average value and standard deviation. Similar statistics on weight and amount of material used.

(iv) Economic data—statistics on material cost per unit. Also it is easily possible to monitor the power consumption on each machine so that electricity costs and machine efficiencies can be computed. This will be very useful in order to see the effect of energy-saving measures (guards, accumulators, etc.).

(v) Key parameters—again the choice of these parameters will vary, but typically will be: set-up time; cavity pressure (may be computed from hydraulic pressure); melt temperature (may be selected barrel temperature); tool temperature—moving half and fixed half; holding pressure and fill, hold and gate times.

(vi) Alarm conditions—it is usually arranged to flash the appropriate parameter in alarm on the VDU.

11.3 Control Strategies

On the basis of information gathered, it is possible to automatically adjust the machine parameters, for example, if the average weight of mouldings is too high, the injection stroke can be adjusted automatically; plastication times, fill times, etc., may be similarly corrected. These strategies are readily programmed in and therefore make the system very powerful. In addition, more advanced control schemes of an adaptive nature can be implemented. Some of these will now be considered.

(i) Adaptive shot size control—the size of each shot is controlled so that the actual cushion of plastic remaining after the packing time is over is equal to the operator-entered cushion reference dimension (the velocity/pressure profile controller working on trip pressure). This is achieved by comparing the actual cushion with the desired cushion and correcting the next shot by adjusting the fully retracted position.

(ii) Adaptive melt control—difficulties in designing a reliable melt temperature transducer have already been covered. However, reasonably accurate measurements have been achieved by devices in the melt, deep in the barrel or at the screw tip. Melt temperature can be corrected quickly by altering back pressure and/or screw speed; operating subsequently on barrel temperatures will bring the back pressure and screw speed to their original values (limits on both these parameters will be necessary).

(iii) Adaptive viscosity control—viscosity control requires the fitting of melt pressure transducers. Then, since nozzle and runner restrictions are fixed, viscosity is calculated from injection speed and melt pressure. Viscosity can then be controlled by, say, varying the barrel temperatures.

12 FUTURE DEVELOPMENTS

Before considering future developments, it is worth while restating the effect of a distributed intelligence approach on instrumenting an injection moulding shop. Instrumentation for each machine consists of several units, split by function, all capable of intercommunicating and of linking to a central computer. The function of this computer is specifically management reporting, data storage and some higher level control strategies.

Controls on each machine can be minimal but will largely be dictated by shop-floor practices. In future, it is likely that each machine will be equipped with a simple operator's control with only function-select manual pushbuttons and fault diagnosis switches. Some form of visual display is necessary, if only for alarm conditions. With the decreasing price of intelligent VDUs (distributed intelligence again) it is highly likely that these will ultimately be fitted to each machine. They could, for example, display velocity, pressure profiles during injection, and also flash alarm messages under fault conditions.

Entry of parameters at the machine could be through a panel of the type discussed in Section 10, but, in fact, this could be condensed down to a hand-held calculator-type device used only during commissioning of a new tool. This is already technically feasible; it is simply a case of establishing practices to suit the new philosophy.

Standardisation of bus structures is essential to give the machinery manufacturer and the end user multi-sourcing of components. The HPIB parallel bus is probably not right in this context but the serial RS232 undoubtedly is. The only question mark relates to its data handling capacity. However, with the approaches outlined, this should not be a real worry.

It is now worthwhile considering the various areas in connection with future developments. Firstly, temperature control. The most important developments have been in the recognition and the achievement of accurate control using deep, properly sited thermocouples; this will undoubtedly continue both with heat and with heat–cool systems. Similarly, the need to accurately control parts of the machine other than the barrel—hot runners, mould, hopper air, etc.—will become increasingly recognised. There is still the need for a reliable, accurate melt thermocouple which is easily mounted, and there are some promising developments in this area. These involve applying heat to the back of the device to eliminate gradient errors.

The need for velocity/pressure profile control on the injection unit will become increasingly recognised. As well as all the advantages outlined previously in respect to moulding quality, the forces on the press can be significantly reduced. Increasingly, such units will be applied to the more complex mould units because of the ease of set-up. To take full advantage of a velocity/pressure profiler on the injection unit will usually involve changes to the hydraulics. Best results have been achieved not with the system outlined but with the system shown in Fig. 28. This has accumulators and

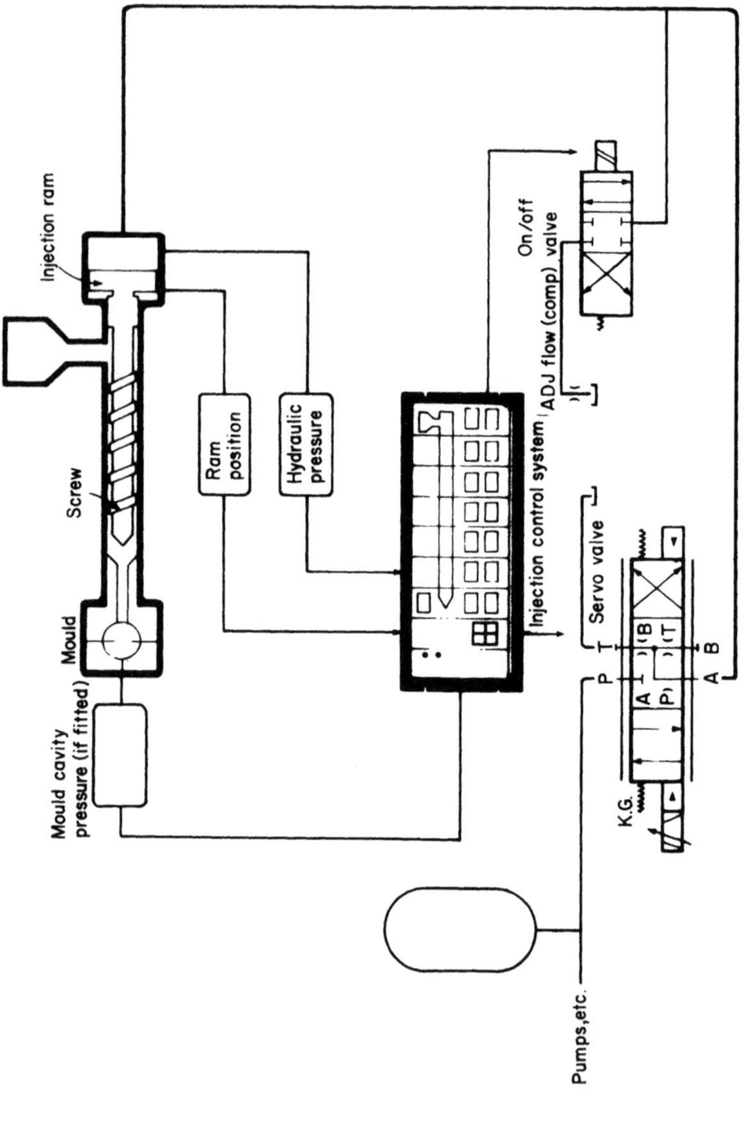

FIG. 28. Improved hydraulic system.

both pressure and velocity profiling are achieved on the servo valve. This allows extremely fast, tight control. Accumulators have, of course, many advantages in energy saving and require smaller motors for a given machine. They, too, are being more widely fitted.

There is still a lot of experimental work necessary to determine the effect of profiling other parameters (hold-on pressure, screw speed, etc.), which is only possible as the techniques outlined are applied. In effect, this is following the principle of tight, local control on all machine parameters in order to minimise interactive effects which, on a badly controlled machine, make it impossible to relate cause and effect.

One of the most promising areas of more sophisticated control is constant shot volume control. The simple technique of varying the fully retracted position is clearly not optimum and further investigation is necessary here.

Obviously there will be significant improvements in production control and quality assurance, made possible by the techniques outlined. Much needs to be done on devices for automatic quality checking—auto-weighing, surface finish scanning, etc. All in all, the future is bright and exciting with much to be done but with significant rewards to be grasped.

INDEX

GPSR Compliance

The European Union's (EU) General Product Safety Regulation (GPSR) is a set of rules that requires consumer products to be safe and our obligations to ensure this.

If you have any concerns about our products, you can contact us on ProductSafety@springernature.com

In case Publisher is established outside the EU, the EU authorized representative is:

Springer Nature Customer Service Center GmbH
Europaplatz 3
69115 Heidelberg, Germany

Batch number: 10607194

Printed by Printforce, the Netherlands